QUALITY AND PRODUCTIVITY FOR BANKERS AND FINANCIAL MANAGERS

QUALITY AND RELIABILITY

A Series Edited by

Edward G. Schilling

Center for Quality and Applied Statistics
Rochester Institute of Technology
Rochester, New York

1. Designing for Minimal Maintenance Expense: The Practical Application of Reliability and Maintainability, *Marvin A. Moss*
2. Quality Control for Profit, Second Edition, Revised and Expanded, *Ronald H. Lester, Norbert L. Enrick, and Harry E. Mottley, Jr.*
3. QCPAC: Statistical Quality Control on the IBM PC, *Steven M. Zimmerman and Leo M. Conrad*
4. Quality by Experimental Design, *Thomas B. Barker*
5. Applications of Quality Control in the Service Industry, *A. C. Rosander*
6. Integrated Product Testing and Evaluating: A Systems Approach to Improve Reliability and Quality, Revised Edition, *Harold L. Gilmore and Herbert C. Schwartz*
7. Quality Management Handbook, *edited by Loren Walsh, Ralph Wurster, and Raymond J. Kimber*
8. Statistical Process Control: A Guide for Implementation, *Roger W. Berger and Thomas Hart*
9. Quality Circles: Selected Readings, *edited by Roger W. Berger and David L. Shores*
10. Quality and Productivity for Bankers and Financial Managers, *William J. Latzko*

Additional volumes in preparation

QUALITY AND PRODUCTIVITY FOR BANKERS AND FINANCIAL MANAGERS

William J. Latzko
Latzko Associates
North Bergen, New Jersey

MARCEL DEKKER, INC.
NEW YORK • BASEL

ASQC QUALITY PRESS
MILWAUKEE

About the Artist: David M. Saunders works for ARBOR, Inc., Philadelphia, providing training and consultation on quality improvement teams for banks, government agencies, utilities and manufacturers. He has also worked with the author, William J. Latzko, on the application of statistical methods in the workplace.

Library of Congress Cataloging in Publication Data

Latzko, William J., [date]
Quality and productivity for bankers and financial managers.

(Quality and reliability ; 10)
Bibliography: p.
Includes index.
1. Banks and banking—Quality control. 2. Service industries—Quality control. I. Title. II. Series.
HG1616.Q34L38 1986 332.1'068-5 86-13572
ISBN 0-8247-7682-8

MARCEL DEKKER, INC.
270 Madison Avenue, New York, New York 10016

Current printing (last digit):
10 9 8 7 6 5 4 3 2 1

PRINTED IN THE UNITED STATES OF AMERICA

Preface

This text is the first definitive book on the topic of quality control in banking. But it goes beyond that. Banking is a subset of the service industry. Insurance, government, transportation and office work in manufacturing are a few places where the same methods have been used.

This book stresses the importance of top management's participation. Top management must allocate time to assure that the principles leading to quality are in fact practiced in their banks. This cannot be delegated with any degree of success. It means that a great deal of time is needed in the beginning. To be successful, top management must allocate as much time to quality as to budgeting, lending and asset-liability management. This is a big step to take. I have seen a number of top managers begin a program with great enthusiasm. Some early results deceived them into thinking the job was done. They no longer paid attention to the process they started. The program failed.

The formula for success requires patience and tenacity on the part of top management. It takes several years for a program to move from the arena of being just "a bright idea that the employees can outlast" to being a way of life. There are no lasting "partial" benefits. Quality is achievable only from the top down. Any program that does not have this type of support is frivolous and will fail. There are enough examples of false starts to be found among the top 20 banks to support this position.

Application of quality management to operations alone does not guarantee success. Continental Illinois had two great practitioners, Larry Eldrige and Chuck Aubrey. They did very fine work in the operations area. Continental did not fail due to their operations. Their failure was due to lending problems compounded by a lack of confidence on the part of the public. I feel certain that if Eldrige and Aubrey had been given a bank-wide charter, Continental's failure would have been averted.

Will the First National Bank of Boston survive its current adverse publicity? They were recently fined $500,000 for failing to report transactions that the law required them to report. Failure on the part of employees to do the right thing—the ultimate responsibility must rest with top management. If they had a mechanism to assure that employees knew their jobs and supported this with a control chart, I believe that they could have avoided their problem.

Every day we read of major banks having massive write-offs due to bad loans or other problems. Could they have been avoided? I think the answer is yes. Bank executives certainly hope that such losses are avoidable. Unfortunately, they rely on inspection and inspection is not reliable. To have proper control a bank must use modern tools. These are the little-known tools of quality management.

If problems in the banking industry persist, banks may well find themselves faced with a government-imposed requirement to control their quality. They will find this as distasteful and costly as the pharmaceutical industry finds the government-imposed "Good Manufacturing Practices" (GMP) and "Good Laboratory Practices" (GLP). Are "Good Banking Practices" (GBP) requirements around the corner? Banks are well advised to put their houses in order before this happens.

The business of banking is lending. A bank is a money store that buys and sells a commodity: money. The buying is the liability side (deposits, certificates of deposits, other money market instruments and borrowings). The selling is the asset side (lending). The essence of lending is getting repayment.

Banking transactions are made on the basis of financial acumen and trust. A major part of lending is assessing whether a borrower can generate enough funds using the proceeds of the loan to repay the bank, pay interest and make a profit. No one would lend if this were not the case.

What would prevent a corporate borrower from repaying the loan? The inability to generate enough funds over the life of the loan. This inability, in turn, is caused by loss of (or failure to gain) market shares. Competition is an important factor—both national and international competition. Bankers making loans would be well advised to determine that the borrower has a quality management system to ward off competition. Many industries are experiencing competition from Japan and other countries that can produce cheaper and better goods. Loan officers are well advised to learn how to tell if their prospective borrower can meet such competition, for example, by following Dr. Deming's 14 points.

Bankers and nonbankers will find in this book a number of topics which cover the technical and the philosophical aspects of quality management. The book has been written for the nontechnical person and contains suggestions on what to do without going into the statistics needed to do the job. Any competent statistician can help the reader with that portion of the implementation.

A book of this sort cannot be prepared without a great deal of support. While I learned from many teachers and colleagues whose principal contributions are mentioned in the appropriate places in the book, I am most grateful to my friend Dr. W. Edwards Deming, who with great patience taught me quality control and the principles of quality management. I have had the opportunity of putting them into practice and seeing them work as well in this country as they have in Japan.

Many people helped with the production of the book. I doubt that a book can be written without the aid and support of the author's family and thanks is due them. My daughter, Victoria Bone, copied substantial portions from an earlier draft to the word processor. My son, Alexander Latzko, provided technical support in the complexity of computers, while my wife, Constance B. Latzko, not only encouraged me but helped in many practical ways. In addition, the help of Jeanine L. Lau of ASQC Quality Press, who in her nice way kept after me to finish the manuscript, and the staff of the publisher, Marcel Dekker, Inc., especially Melodie Wertelet, are greatly appreciated.

A particular debt of gratitude is due to my friend and colleague David M. Saunders, Senior Research Associate at ARBOR, whose fine illustrations bring to life many of the important issues of the text.

William J. Latzko

Contents

QUALITY AND PRODUCTIVITY FOR BANKERS AND FINANCIAL MANAGERS

1

Introduction

Do banks need quality control? Most bank executives would say yes. But they might not agree on the nature or intensity of the quality control system that is best for them. They know that in clerical operations errors frequently can happen. Managers know that these errors can have severe consequences. Some try to cope with this by relying on the techniques which worked in the past. However, they find that the advances in the industry make these methods inefficient and costly. Others hope that automation is a solution to their problems. It becomes rapidly clear to them that while automation solves some known problems, it also creates a host of new conditions causing errors.

What kind of quality control a bank needs is dictated by the changes which the banking industry has experienced recently. These changes coincide with the advent of increased computer usage. While the early uses of this powerful tool related to "its check processing and bookkeeping functions, the industry is also looking at the computer as a source of improved efficiency and useful information" [1]. This automation of bank operations has brought with it the need for new and improved quality control. The quality control techniques of the past are no long practicable or cost effective.

The real question facing bankers today is not so much whether they need quality control—legal and economic considerations make the need clear—but rather in what way to provide quality control. To develop this properly, it is necessary to examine the historical development of quality control in banking and the nature of modern quality control.

The Development of Quality Control in Banks

Quality control in banking is not new. As long as bankers are subject to regulations and external supervision, a control of the quality of their services is needed.

Ancient Laws

The earliest laws of banking which have survived are from the kingdom of Eshunnana (about 2000 B.C.). They dealt with interest rates, the type of investments permitted and penalties. The interest rate was 20 percent, presumably for any length of usage. Section 14 of the code said: "The fee of a (money-lender) should he bring five shekels the fee is one shekel of silver; . . . " [2].

The Code of Hammurabi—The Law of Eshunnana is fragmentary and does not clearly specify the penalties invoked for failing to observe the rules. The Code of Hammurabi (about 1730 B.C.) is far more complete. It also specifies a 20 percent interest rate but gives penalties for malfeasance.

> If a merchant lent grain or money at interest and when he lent (it) at interest he paid out the money by the small weight and the grain by the small measure, but when he got (it) back he got the money by the [large] weight (and) the grain by the large measure, [that merchant shall forfeit] whatever he lent [3].

Other sections in the Code specify double indemnity and in some instances the death penalty is invoked. Under such a law it was left to the banker to control the quality of his service.

The Roman Era—While the earliest laws gave impetus to the banker to check his work, it was up to his customer to complain to the authorities if the banker failed to perform according to law. The Romans appear to have been the first to institute an external audit.

> Specialists in coinage in Rome were known as "Argenterii" or dealers in silver. According to history, under Roman laws, the Argenterii were required to keep cash books, day books, and deposit books, and their operations were subject to official inspection, somewhat similar, no doubt, to the examination of banks made today by banking authorities [4].

The Byzantine method of quality control—The Roman method of auditing the books of bankers remained in effect until the rise of the Byzantine Empire nearly 1000 years later. The Byzantine Empire was built upon the systems concept: every action in the Empire was regulated by procedures which had to be followed to the letter. To enforce these procedures, the prefect (local governor) had attached to his court retinue an official inspector, a *Logothete*. The *Logothete* was charged with the inspection of all workshops and operations performed in the district. If such an inspection disclosed an infraction of the rules, the *Logothete* would have the culprit brought to trial. "Even the lightest punishment was severe and the urban prefect was pitiless. The goldsmith who had taken delivery of more than the permitted pound of unminted gold was fined the equivalent of [$800]" [5].

In addition to the pecuniary punishments meted out, additional severe penalties were inflicted if the transgression was due to more than an oversight. "'Every goldsmith or banker who buys unminted gold for his own use shall have a hand cut off,' said the edict" [5].

A system of quality control inspectors and stern sanctions kept the Byzantine Empire intact for a thousand years. In fact, this form of quality control has survived into this century. While the sanctions are not as harsh as they once were, many banks still rely on a checker to catch those who make mistakes so that "appropriate" action can be taken.

Modern Quality Control

The need for quality in banking is as great today as ever before. Not only do bankers have to live up to a host of government regulations but the marketplace requires a basically fault-free service. Robert McCreary, in discussing the topic, stated: "We are living today in a world where communications about quality, consumer awareness and comparative buying are major factors in supporting a price for good quality in a product" [6]. The consumer has a great desire for quality. A recent Citicorp Survey showed that consumers were displeased with both the pricing and quality of services. However, lowering price was not the answer:

> Despite the obvious displeasure over pricing, people are not willing to trade off quality for a better bargain. More than three out of every four in the poll said quality is "much more important" or "slightly more important" than price [7].

Automation, while aiding banks in performing their work faster and better also increases the risk of errors. Mistakes do not occur as often as under the manual systems, but when they do happen they are almost always major and occur in massive amounts. The story of Key City Bank and Trust Company reported in the New York *Times* is a case in point (see Figure 1.1). Under the title "Dubuque Bank Gets a Little Too Friendly," the *Times* stated:

> "Let's be friends" is the slogan of the [bank], but some customers of Dubuque's smallest commercial bank were a bit surprised when they found out how far that slogan went.
>
> Instead of receiving their normal Christmas Club checks last week, several customers got checks for more than $1 million.
>
> "I guess our staff is so imbued with being the friendliest bank in town, we got a little carried away," Melvin Murrack, president of the bank, said. He said a clerical error in the processing resulted in the incorrect checks [8].

With automation mistakes do not occur as often as with manual systems, but when they do they are almost always major and occur in massive amounts.

Figure 1.1 Automated mistakes are almost always major and occur in massive amounts.

In the past eight years a number of banks have recognized the increased demand for quality and the need for modern control procedures to match the advances obtained by the computerization of bank services. At least one bank has advertised its quality control program. In a supplement to *American Banker*, Chemical Bank stated, "What we need from a New York bank is fast, responsive service and a lower error rate. That's the whole point of Chemical's quality control program" [9].

The Nature of Modern Quality Control

To determine the nature of modern quality control it is important to examine the concept of quality control, the basis for the modern theory of quality control, the tools used and the economics of using these methods.

What Is Quality Control?

The two words "quality" and "control" are significant in describing the concept. Of these "control" is better known and easier to describe.

Definition of control—Control implies fluctuations. In every operation there is a degree of variability which when kept within limits is said to be controlled. As Dr. Shewhart stated in his basic book on quality control:

> For our present purpose a phenomenon will be said to be controlled when, through the use of past experience, we can predict, at least within limits, how the phenomenon may be expected to vary in the future. Here it is understood that prediction within limits means that we can state, at least approximately, the probability that the observed phenomenon will fall within given limits [10].

In terms of banking operations this can be interpreted to mean that if the reject rate of a MICR Reader/Sorter is between 0.8% and 1.0% for a given volume and type of work, the process is in control if all the observations fall within that range of rejects. Or, if in a clerical operation 2.5% plus or minus 0.5% of the items are not prepared correctly, the process is said to be in control if the vast majority of the work sampled falls within an range of 2–3% defective items in the sample.

Definition of quality—"Dating at least from the time of Aristotle, there has been some tendency to conceive of quality as indicating the *goodness* of an object" [11]. There is, however, an inherent difficulty in this definition. Both "quality" and "goodness" are transcendental terms which cannot be defined; at best they can be described. What is quality to one person is not necessarily quality to another person. As R. D. Sonderup states it:

> Quality means many things to many people. To a salesman, it is a magic word to be used as many times as possible in his sales presentation. To a company president, it is a reputation that must be achieved and, once achieved, maintained. . . . To the consumer, quality is that property of a product that creates a desire for continued use of ownership. Advertisement, salesmanship, style, feature, and price can all create a desire for *continued* use or ownership. Quality control is, then, a control of those characteristics of the product that create this desire [12].

Shewhart explains that "Quality, in Latin *qualitas*, comes from *qualis*, meaning 'how constituted' and signifies such as the thing really is" [13]. Frank Squires amplifies this by adding, "Actually, 'quality' is not something that can be isolated from the product and controlled. It is the essence of the product—the characteristic for which it is manufactured" [14].

It may help to consider that there are two aspects to quality: the design aspect and the production aspect. Design quality is the quality required to meet the customer's specification. These specifications may be explicit or implicit. The use of standing instructions in an operation is an example of explicit specifications. The use of good banking practices is an example of the use of implicit specifications.

Production quality is the quality of the work as it is produced. If the work is produced according to the customer's specification, it is good; if not, it is defective.

Definition of quality control—"Quality control is the determination that the production quality conforms to the design quality" [15]. This means that the work as it is produced meets the customer's specifications.

Why Use Statistical Methods for Quality Control?

The quality control described so far is the same as that used in the Byzantine Empire. It consists of a checking function together with a corrective action function. It is a very generalized technique. However, as a technique by itself it is very wasteful. In the banking industry it is common to check every item produced 100% or more (if multiple levels of checkers are used). While in some cases this may be the best course of action to take, in many instances it is being done because that is the "traditional" method of controlling quality.

Manufacturing industry followed pretty much the same path until the late 1920s when Dr. Shewhart did his pioneering studies at the Bell Laboratories. Because of a concern with the economics of obtaining quality in production, he and others formulated the statistical methods of quality control used today. These techniques were found to be highly efficient and economical. In some instances, such as in the case of destructive testing, these methods were the only practical solution.

The statistical quality control methods are widely believed to be responsible for the ability of the United States to produce many of the items required in World War II. The use of these methods proved so successful that the government requires its use with all contractors: "This specification requires the establishment of a quality system by the contractor . . . subject to surveillance by the Government representative" [16].

While the underlying theory of the statistical quality control methods are mathematical, most of the work done in executing such a system is clerical in nature, requiring, for the most part, only observing, counting, and tallying.

Quality Control Techniques Applicable to Banking

While nearly all of the methods of statistical quality control are applicable to banking, three major areas are particularly pertinent: acceptance sampling, process controls and special studies.

Acceptance sampling—In a number of areas of bank operation, the quality of the materials used is of the utmost importance. The checks processing and computer operations have particular needs for satisfactory raw materials. Other areas in a bank such as where microfilming is performed and the purchasing function are interested in obtaining satisfactory materials for use. In many cases the test for quality is destructive—microfilm must be exposed to test it. Drawing a random sample of the product and testing the sample, when done according to a proper design, can give assurance that the materials will work as specified.

Process control—By using statistical techniques to collect data and then analyzing this data, the process capability of a system can be determined. Information about the deviations from the process capability can be used by the operating management to effect correction, thereby, achieving the process capability. Once this goal has been obtained, further measurement gives assurance that the quality level is being maintained.

Process control systems are very applicable in checks processing areas as well as in many clerical operations areas. Teller over and short data can be controlled by such methods as well.

Special studies—A number of the methods used for quality control have applicability in solving specific problems sometimes only tangentially related to quality. Teller queuing, testing of new equipment for reliability, sampling a data file for operating information are a few examples of these techniques.

Many additional applications of quality control are possible. Among these are Evolutionary Operations Processing (EVOP) and design of experiments.

How Bankers Go About Achieving Top Quality

There are many necessary preconditions to achieving top quality. These are outlined in the succeeding chapters and deal with the support and participation needed from executive and middle management as well as the staff. Once that this type of interactive commitment

HOW BANKERS CAN GO ABOUT ACHIEVING TOP QUALITY

Step #1 Determine what to measure
□ use QMS (Quality Measurement System)

Step #2 Get system into control
□ use QUIP (The Quality Improvement Program)

Step #3 Improve the system
□ use MBT (Managerial Breakthrough Technique)

Figure 1.2 Three steps to better quality.

is in place, bankers can follow a pattern to success (see Figure 1.2 and Figure 1.3). The three basic steps are

1. Develop what to measure
2. Achieve control
3. Make improvements

Develop what to measure —The method of Quality Measuring System (QMS) is a powerful tool to use for this purpose. It is described in Chapter Eight, "Starting a Quality Program." This method is outstanding in two ways:

1. The measures are readily determined.
2. The method results in enthusiasm and participation of the whole staff in the developing of better quality.

Achieve control—The method used for this purpose in a bank is the Quality Improvement Program (QUIP). Used with the appropriate control charts, this powerful tool allows bankers to bring the system under control and eliminate the special causes of defects. Chapters 4 and 5 deal with this program.

Make improvements—The method found most useful for this purpose is Dr. Juran's Managerial Breakthrough Technique (MBT). The uni-

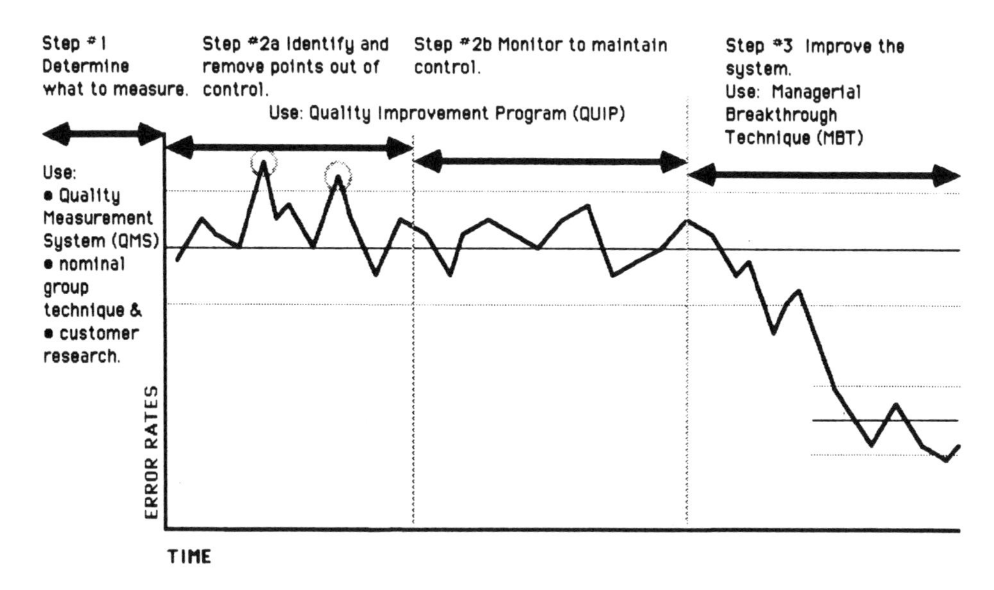

Figure 1.3 Application of the 3 steps.

versal approach of this method has yielded outstanding results where applied. It is briefly covered in Chapter 9.

It should be noted that the sequence of the methods is of importance. Some banks attempted to go directly to the improvement phase and found they were reacting to special causes and thus got very little done. The tool is powerful in the right circumstances, namely, when the system is in statistical control.

A special case of MBT is the quality circle concept. Major problems arise when this is used without the necessary pre-conditions. Lack of management support in a bank wide quality program can stifle such programs. If the opinion is that quality management applies only to clerks in operations, the program has failed. All activities of the bank, chairman, money management, lending, trust operations, marketing, personnel, data processing among others must all be part of the program.

To succeed quality management must be incorporated into the bank's culture and be a way of life.

Summary

Bankers have needed quality control for centuries. This need has changed with time as has the degree of sophistication required to maintain an error free environment. The advent of the computer has intensified the need to determine the best way to provide quality control to an industry whose technology is evolving rapidly. The focus of this book is to examine the ways to incorporate quality control into bank operations.

References

1. Paul S. Nadler, "A Look at the Future of American Banking," in Herbert V. Prochnow and Herbert V. Prochnow, Jr. (Editors), *The Changing World of Banking* (New York: Harper & Row, 1974), p. 385.
2. Albrecht Goetz (Translator), "The Laws of Eshunnana," in James B. Pritchard (Editor), *The Ancient Near East*, Volume 1, 6th printing (Princeton: Princeton University Press, 1973), p. 134.
3. Theophile J. Meek (Translator), "The Code of Hammurabi," in James B. Pritchard (Editor), *The Ancient Near East*, Volume I, 6th printing (Princeton: Princeton University Press, 1973) p. 148.
4. Marshall C. Corns, *The Practical Operation and Management of a Bank*, 2nd edition (Boston: Bankers Publishing Co., 1968), p. 33.
5. Renee Guerdan, *Byzantium, Its Triumphs and Tragedy* (New York: George Allen, 1956), p. 96ff.
6. Robert M. McCreary, "Whence Cometh Quality Control," *Quality Progress*, Volume IX (July 1976), p. 15.
7. "Poll Finds Consumers Unhappy on Quality," *American Banker* (New York), September 22, 1976, p. 7.
8. "Dubuque Bank Gets a Little Too Friendly," New York *Times* November 14, 1976, p. 14.
9. Chemical Bank Advertising Supplement, *American Banker*, November 15, 1976.
10. W. A. Shewhart, *Economic Control of Quality of Manufactured*

Product (Princeton: D. Van Nostrand Company, Inc., 1931), p. 6.

11. W. A. Shewhart, *Economic Control of Quality of Manufactured Product* (Princeton: D. Van Nostrand Company, Inc., 1931), p. 37.
12. R. D. Sonderup, "Quality Control and Product Reliability," in Robert Finley and Henry Ziobro (Editors), *The Manufacturing Man and His Job* (New York: American Management Association, Inc., 1966), p. 175.
13. W. A. Shewhart, *Economic Control of Quality of Manufactured Product* (Princeton: D. Van Nostrand Company, Inc., 1931), p. 38.
14. Frank H. Squires, "Pretty Good Isn't Good Enough," *The Management Review*, Volume 48 (October 1959), p. 18.
15. William J. Latzko, "Quality Control in Banking," in *The 1974 National Operations and Automation Conference Proceedings* (Washington: American Bankers Association), p. 37.
16. "Quality Control System Requirements," *Military Specification MIL-Q-9858* (Washington: Superintendent of Documents, 1956), p. 1.

2
High Speed Data Handling

Bank operations can be compared to a factory process; more specifically, a paperwork factory. Bank operations are responsible for turning out massive volumes of transactions. It has been noted that the production process consists of two distinct types: high speed mechanical processing and low speed clerical processing [1]. The high speed processing closely resembles the processing of manufacturing operations. Because of this, the application of standard methods of statistical quality control, successfully used in factories, were the first to be tried in banking and are in common use today.

A survey performed in 1975 by the Banking Subcommittee of the American Society for Quality Control showed that 73% if the respondent banks had a formal quality control program. Most of these were in the area of checks processing, a high speed, automated operation [2]. Even though the response to the survey was small, and even though the survey was limited, it was interesting to note the amount of response received. The results indicated substantial quality control activity in the high speed operations of the bank. A copy of the survey is reproduced in the appendix.

The use of quality control in automated areas of the bank has grown. In a recent book, Roger Langevin stated, "Most banks now have ongoing quality control programs to prevent MICR problems" [3]. By the word "most" he means the larger banks in the country.

Applicable Bank Operations

The most likely candidates for statistical quality methods used in industry are the automated areas of bank operations. These are the checks processing area, the computer area and the microfilming of bank documents. These are the most common applications today. As the industry continues to mechanize, other areas such as credit cards, funds transfer and trust operations will also be candidates for quality control using the techniques invented for manufacturing.

Checks Processing

Recently, Walter Stafeil of the Bank Administration Institute wrote that "the cost for total check collection services would increase 2.7 times [from 1973 to 1980] while the cost for handling exception items would increase 5.8 times" [4]. To combat the rising expense to banks, the BAI sponsored an Exception Item Conference in March, 1976. At this conference, a proposal was made to reduce exception items by the following recommendations [5]:

1. Regional Interbank Tests such as those performed in New York and Philadelphia should be performed in all regions.
2. Consideration should be given to a standard set of definitions so that regional results can be compared.
3. Individual banks should establish MICR testing. Banks can either establish their own testing procedures or arrange to have documents tested by others.
4. A standard methodology of acceptance sampling should be established. The methodology and standards in this paper [The MICR Challenge for Bankers] are proposed as a suitable means for measuring MICR Quality.
5. More complete standards than those which currently exist should be established for the paper used in Checks Processing.

These recommendations were accepted and a joint task force was established to implement them. The result of this effort was reported by Mr. Stafeil.

> The second recommendation of the working group calls for the adoption and improvement of quality control programs within all banks. Much had been said and little done in this area—few banks actually exercised a strict quality control program.
>
> The working group has urged that this be an area of future work and that a handbook or guideline for the establishment and operation of an effective quality control program should be prepared [4].

The concern for the quality of the checks processing operation is easily understood when one considers the cost involved with MICR rejects. These costs are generally estimated to be from $0.10 to $0.30 per rejected item. In addition to the usual out-of-pocket costs there are a substantial number of hidden costs such as lost float and expenses in the account reconcilement operation of the bank [5]. (See Figure 2.1 and Figure 2.2.)

Since check processing consists of treating a vast number of documents, and since even a small percent of bad work can cost a large amount of absolute dollars, the rate of return in applying quality control in checks processing is high. Table 2.1 gives some idea of the annual cost of MICR rejects. Using 3%, which is approximately the average national reject rate report by the BAI, [6] a 250-day work-year, and a 20 cent correction cost, a bank which daily processes some 300,000 checks spends $450,000 per year to correct MICR rejects. A quality control effort which can reduce the rate to 2% will save this bank $150,000 per year.

Statistical quality control is quite capable of achieving major cost savings by reducing MICR rejects at a relatively small investment. The combination of large volume and mechanical methods of processing allow the efficient use of these techniques.

The check processing function starts with the purchase of checks and other internal MICR forms used for processing. The first area for achieving good quality is, therefore, the purchasing department. The use of proper purchasing techniques, vendor qualification, lot ac-

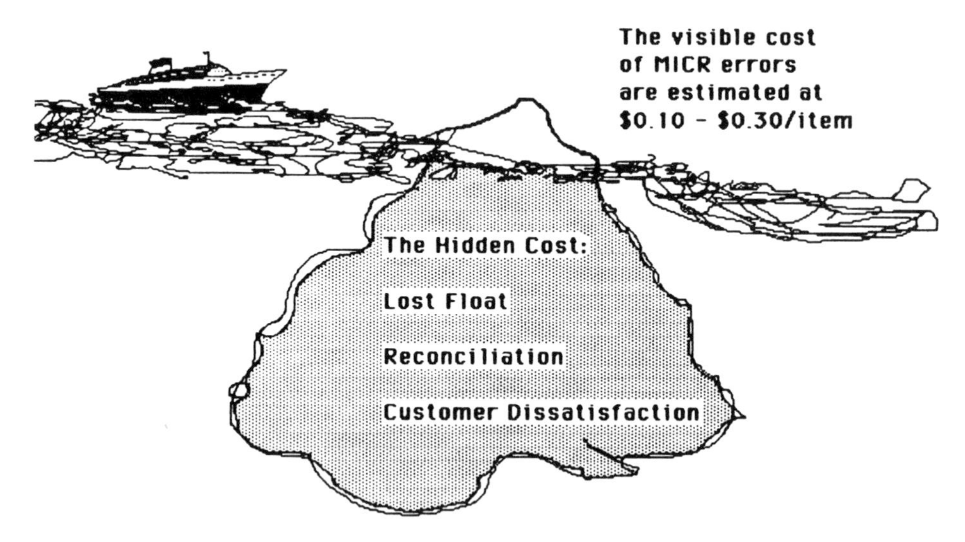

Figure 2.1 The hidden costs of MICR rejects.

ceptance sampling and vendor evaluation are primary techniques found useful in this area.

The second place where quality control is needed is in the encoding process. In the department where the dollar amount and, sometimes, other information is placed on the document, quality is of the utmost importance. The techniques of set-up tests and patrol inspection have been found to work best here.

TABLE 2.1 Annual Costs of MICR Rejects at Given Reject Rates

Daily Volume of Checks	MICR Reject Rate 2%	3%	4%
100,000	$100,000	$150,000	$200,000
200,000	200,000	300,000	400,000
300,000	300,000	450,000	600,000
400,000	400,000	600,000	800,000

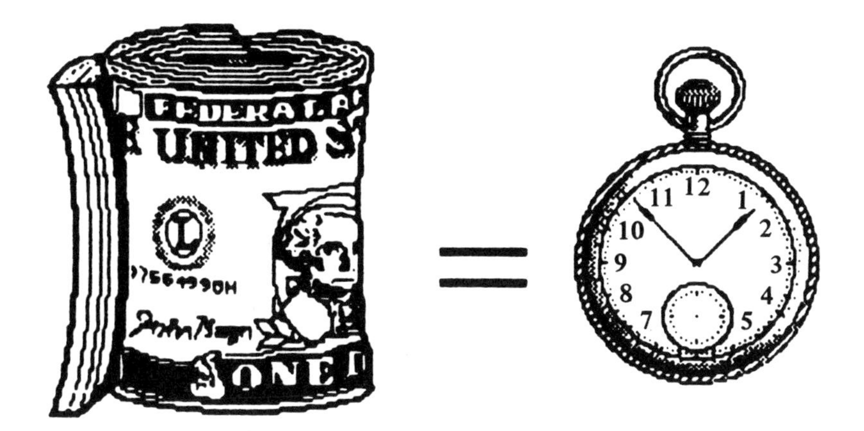

Figure 2.2 Time equals money and money equals time.

The third major point of control is in the reader/sorter operation. Here the bank depends on the customer engineer of the manufacturer to properly maintain the equipment. Furthermore, the operator of the equipment plays a vital role in the proper functioning of the system. The forms of control found most useful at this point are control charts and records which permit the evaluation of the reliability of the equipment. In the technical sense reliability is the ability of a system to perform its intended operation for a given period of time under given conditions. In the case of reader/sorters it is the ability of the equipment to read good MICR characters as presented.

Computer Processing

Check processing is a special case of computer processing. Since it is of such paramount importance to the banking industry, it has been covered in somewhat more detail.

Computer processing also consists of four phases requiring quality control. These are

1. Raw materials
2. Input preparation
3. Processing
4. Output

In fact all areas of high speed data handling have these four aspects requiring adequate control to assure the desired quality of the final product.

Raw materials—The raw materials of computer processing are the cards, tapes, disks, forms and other input/output media used in the operation of the system. It could also properly include the computer library operations since the proper control of tapes and disks is essential to the quality of an ongoing system.

The techniques used range from simple logs, spot checks and like elementary control techniques to the more sophisticated methods of acceptance sampling, data inscription control and check routines.

Input preparation—A good deal of work has been done in this area. The Bureau of the Census has been particularly active in this field. Recognizing that erroneous input can have drastic effects on Congress, the Bureau has been in the forefront in developing quality control techniques to minimize the input error. While most of their work has followed standard techniques of acceptance sampling, one rather unique method was proposed for the control of quality of the keypunching of certain data cards. This technique is an application of game theory to the preparation of data cards. In effect, qualified operators are given ten points when they start the job of keypunching. Samples of their work are verified. For every such lot accepted the operator is given an additional point up to a maximum of 20 points. For every lot rejected, the operator loses a point. If an operator exhausts all points, the operator is taken off the job [7].

Input control ranges from none to keypunch verification (not as reliable as many people think) [8] to more sophisticated statistical controls as noted above.

Processing—There are two aspects to processing:

1. The actual operations of the computer
2. The programs which are used to achieve the desired results

For the actual operation control, a number of methods have been successfully applied. These range from supervisory control to the use of control charts. Since the physical operation of a computer is essentially clerical, this aspect is best discussed under the chapter dealing with clerical control.

The control of programming poses an entirely different problem. The building of systems and particularly their modification can often lead to a variety of serious but hard to detect error conditions. To combat this situation several approaches has been proposed. Among these is the use of a technique known as "Decision Tables," which allows the programmer to follow every one of the many paths that a decision can take and, thereby, spot any impacted but unforseen area [9]. Other methods are described in Eugene Kirby's paper on "Quality Control in Banking" [10].

Output—Very little is done with regard to the control of the output. Normally, it is felt that if the input and the process are controlled that the output will automatically be in control. While this may be generally true, some banks do have an output quality control in the same way that clerical operations have a final verification. Normally, these operations include a control of the logic produced and the accuracy of the distribution of the final product. In some installations, control of the timeliness is also included. In a bank, computer quality means not only accuracy but also timeliness.

Other Operations

As automation progresses it can be expected that more and more of the bank's operations will change from clerical to high speed type of work. Most of these areas are computer connected, either to a mainframe, central unit or a stand-alone minicomputer. The nature of the hook-up is not material; the nature of the work determines whether it is high speed or clerical. If the operation is essentially mechanical, the techniques of high speed data processing apply. One application found in most banks is that of microfilming. Most vital records of the bank are microfilmed in one of three modes:

1. Planetary
2. Rotary
3. Computer Output Microfilm (COM)

Most banks use rotary microfilming either on-line or off-line to microfilm checks and related documents. A number of banks develop their own film and some even duplicate these films. The final output is achieved from microfilm readers some of which can also produce a photocopy of the screen image.

Microfilm processing certainly qualifies for high speed data handling. From the raw material, (film, chemicals and input documents) to input preparation (exposure, speed of filming and indexing) to the processing and finally the output (reading), the process requires first class control to obtain usable results. If taxpayer records are involved or legal uses are foreseen, the Internal Revenue Service has specific quality requirements which must be met [11]. For other use including archival data, the procedures outlined by the American National Standards Institute (ANSI) and the National Microfilm Association (NMA) should be observed [12]. For those who process their own film, the Kodak Corporation has prepared a useful brochure on microfilm quality control [13]. Regretfully, too few banks are aware of these requirements and pay attention to them.

As storage requirements, both legal and the bank's own needs increase, the need for more reliable microfilm will focus attention on this area of the operations. The elimination of paper transactions in banking will also cause a shift to microfilm. It will be prudent bank management to have high reliability in this area.

Useful Techniques

A number of methods for controlling the quality of high speed data handling operations have been mentioned in passing. These ranged from the concept of *laissez-faire* to statistical methodology. The proper choice will depend on the method which is most cost effective. In some cases the use of the more sophisticated methods can be a matter of overkill.

Most managers are amply acquainted with the technology of their operations and elementary methods of checking that quality has been achieved. There are two general methods of quality control which are extremely useful but not as well known as they deserve to be. These methods are those of sampling, particularly acceptance sampling, and process controls. Some considerations are needed in choosing the most appropriate method.

Acceptance Sampling

Sampling is the taking of a portion of a larger body and estimating some characteristic of the larger body from observations made on the portion. In making the estimate two possible errors exist:

1. A bias because the portion was not representative
2. Inaccuracy because the portion was too small to make a useful measurement

The bias can be greatly reduced and even eliminated by proper sampling procedures. The inaccuracy can be reduced by taking enough sample to make the result precise enough for practical purposes. Any professional statistician or quality control engineer can determine the proper method and sample size to achieve the purpose of the sample.

In many cases the object of the sample is not to estimate some characteristic but merely to determine that a batch of material meets some minimum (or maximum) requirement. In such an instance, a special form of sampling called "acceptance sampling" can be used. Acceptance sampling is used so often in quality control that special tables have been prepared to assist the general practitioner in selecting the appropriate sample size as well as the proper method to obtain the desired results.

While some acceptance sampling tables, such as MIL-STD-414 [14], deal with variables (i.e., some continuous measurement such as dollar amount), most of the tables deal with the far more common characteristic, attribute measurement. An attribute measurement is the presence or absence of a characteristic to be measured [5]. For instance, a MICR number meets or does not meet the requirements of the ABA standards. If it meets the requirement, the attribute of the number is "good," otherwise the attribute is "bad."

One such attribute plan which is widely used in banking is a Military Standard, MIL-STD-105D [15]. This plan is easily obtained. With a little care, it is easy to use.

For proper use, the plan requires careful study of the tables it contains. Given a specific lot size, the plan directs the user to a lettered section. This section contains data to select the correct sample size and acceptance number to be applied. For instance, to test a lot of 35,000 checks (see Figure 2.3), Table I of the plan shows sample size code

ACCEPTANCE SAMPLING IS OFTEN USED IN QUALITY CONTROL

A. Start with 35,000 checks

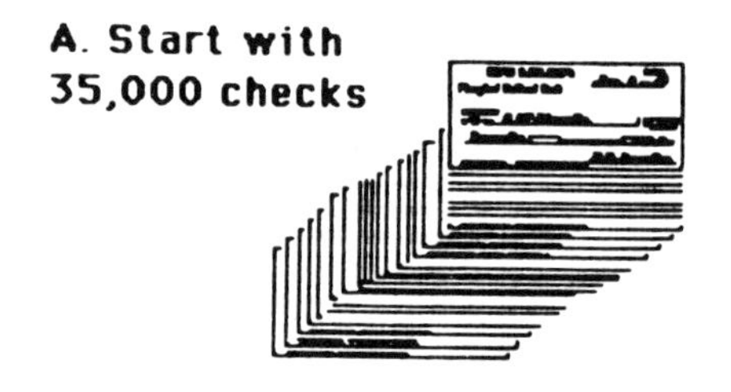

B. Go to book: Military Standard 105D Table I, for General Inspection Level I.

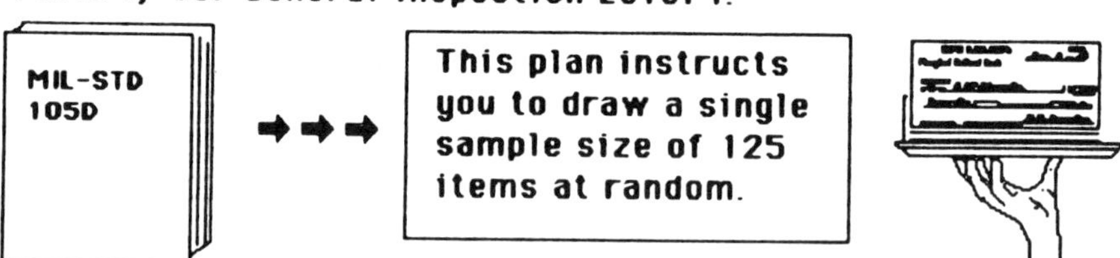

C. Draw random sample of 125 checks and inspect each check.

Decision rule:

Pass all 35,000 checks if only one item fails.

Reject all 35,000 checks if two or more items are defective.

Figure 2.3 How to inspect 35,000 checks.

letter K to be appropriate for General Inspection Level I. Plan K shows that a single sample size of 125 items drawn at random is sufficient for testing. Using an Acceptable Quality Level (AQL) of 0.40% one sees that the lot is acceptable if all 125 items pass inspection or if only one item fails. If two or more items of the 125 are defective, the lot has to be rejected.

Such plans carry a small risk for the consumer, the bank, that an undesirable lot will still be accepted on the basis of the sample. There is also a small risk on the part of the producer, the printer, that an otherwise good lot will be rejected on the basis of the sample evidence. On the whole, these plans tend to balance the two risks on the economics involved. This balance is achieved by the careful selection of the General Inspection Level and the AQL. Past history and cost factors are used to set the optimal values. The values used in the illustration are those which have been found most useful in testing MICR documents [5].

Acceptance sampling is relatively easy. Once the correct use of the tables has been mastered, it will be found to be a highly efficient technique. The question of the sampling method will be covered in more detail in the next chapter.

Process Controls

Acceptance sampling is most helpful when batches of work can be tested to arrive at a decision of whether they are useful or not. It has its place in testing raw materials and like accumulations of work. It is not efficient in controlling a stream of ongoing operations. While it is possible in theory to accumulate the output of an operation and test pseudo lots thus formed, this is a very inefficient use of the method. It is far more desirable to have ongoing or process controls. Process controls are simply periodic tests which monitor the operation as it is being performed and signal any out-of-control conditions. To be able to do so implies that the system has a process capability which is known and can be measured. This capability is subject to some random fluctuation which, if exceeded, signals that something is wrong with the operation that requires immediate attention of the supervision.

If, by way of illustration, the reject rate of a MICR process is two percent plus or minus one-half percent, then the system should signal a problem if the reject rate goes to three percent, or anything over two and one-half percent.

The best way to illustrate the current level of control is with a control chart. (See Figure 2.4.) The best known of these charts is the Shewhart chart. The distinguishing features of this chart are that the abscissa is always a function of time (hours, days, weeks, units in order of production). The ordinate is the variable which is measured. The process capability is shown by the central line enclosed by the upper

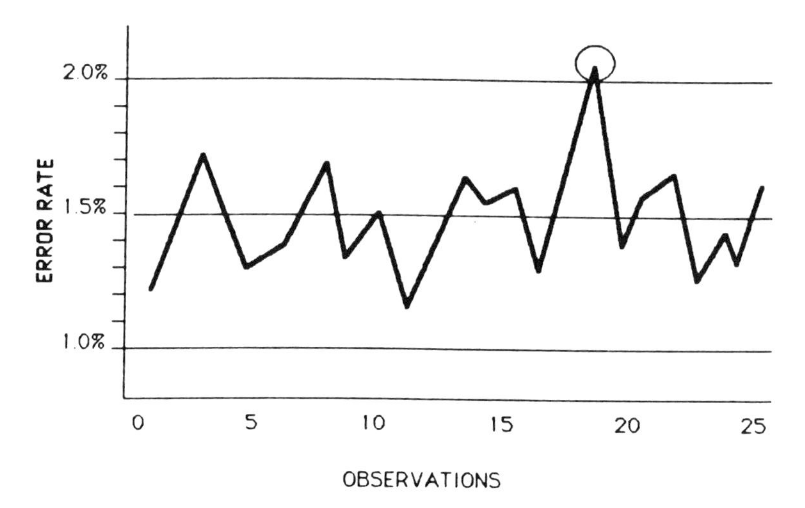

Figure 2.4 Example of a Shewhart Chart.

and lower control limits. Points are plotted sequentially. Any point falling outside the limits is a signal of the existence of an assignable cause that requires resolution.

There are a variety of Shewhart charts as well as other forms of control charts which can be used. The purpose here is not to describe these charts exhaustively but to merely introduce the most important concepts so that a later evaluation of the various techniques available will be more meaningful.

In addition to the use of control charts, process control also uses set-up checks as a method of controlling quality. These are tests made on the equipment about to be used to determine that the machinery functions as intended. As with all mechanical and electronic devices, it is necessary to test them after even short periods of disuse since the mechanisms sometimes degrade.

In the same manner as set-up checks, patrol inspections of the equipment used are frequently valuable. These are tests performed periodically while the equipment is being used. Inspections of this type are useful in that they can detect problems more rapidly than other methods and thereby prevent either more serious damage or long downtimes.

Measuring Process Capability

Without knowing the performance or process capability of their equipment, most managers either cannot control their operations or they over or under control them. MICR rejects are a good example. Many check processing managers do not really know their reject rate other than in qualitative terms such as, "it is average." With this lack of knowledge, problems can be undetected for a long time until they become so severe as to force their attention on the manager. (See Figure 2.5.)

Any form of quality control requires at a minimum that the current process average be known at all times and that the process capability be determined. "The process capability is the lowest error rate attainable from a system under normal circumstance" [16]. If the process average is equal to the process capability then improvement in quality can only be made by changing the system. If the process average is not equal to the process capability, improvements in quality are possible by isolating causes for error and correcting them.

The process capability is measured by the statistical evaluation of historical data. The process is an iterative one. It consists of statistically testing each value against an assumed mean and removing identified outliers to recompute the mean with the remaining values and to repeat this operation until no further outliers exist [16].

Once determined, the process capability can be evaluated by management. If the quality represented by it is adequate, it becomes the target or goal to be achieved. If it is inadequate, systems work is required to improve it.

Control and Reports

Speed and simplicity are the keywords to getting proper measures of control. This means that data must be timely and easy to understand. The Shewhart chart has been found to be an excellent tool for this purpose.

If the data is updated within one period of the value of the abscissa, this type of chart can pay off handsomely by signalling potential problems before they assume too serious an aspect.

This manager takes periodic looks
at his system, but doesn't see the
shaded areas because he uses numbers
but not control charts.

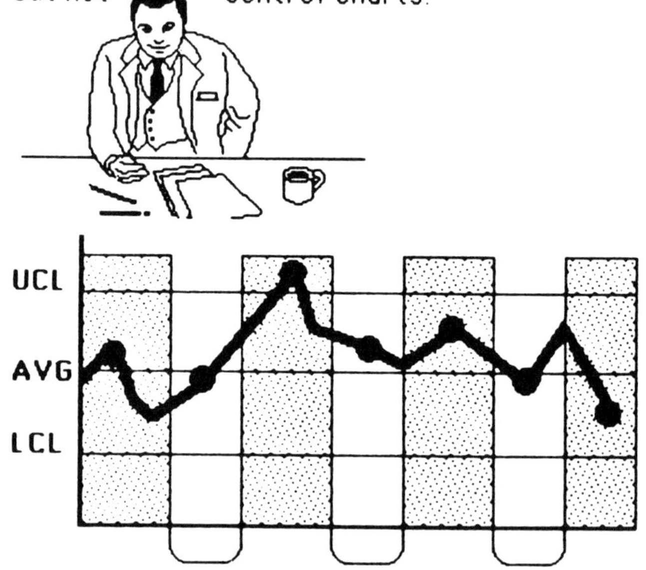

Control lines are set at 3 standard
deviations above and below a well-
established average.
UCL is the upper control line.
LCL is the lower control line.

Figure 2.5 What you don't see can hurt you.

Regardless of the technique employed, controls must be simple and easy to understand while reports used for control must be timely.

References

1. William J. Latzko, "QUIP—The Quality Improvement Program," in *Twenty-Ninth Annual Technical Conference Proceedings* (San Diego: American Society for Quality Control, 1975), p. 247.
2. Eugene Kirby, "Results of a Recent Survey of Banks," unpub-

lished report to the Banking Subcommittee of the American Society for Quality Control, 1975.
3. Roger G. Langevin, *Quality Control in the Service Industry*, (New York: AMACOM, 1977) p. 32.
4. Walter W. Stafeil, "Exception Item 'Horror Story' May Yet Have a Happy Ending," *American Banker* (New York), May 16, 1977, p. 8.
5. From William J. Latzko, "The MICR Challenge for Bankers," a speech presented at the Bank Administration Institute's Exception Item Conference, Chicago, March, 1976.
6. Walter W. Stafeil, *1974 Survey of the Check Collection System* (Park Ridge, Illinois: Bank Administration Institute 1975), p. 3.
7. William H. Cook, *Sample Verification Plan for Punching CATO Cards* (Washington: U.S. Bureau of Census, November 5, 1957).
8. An unpublished study by William J. Latzko and Frederick Schweitzer at CBS Direct Marketing showed verification of a full alphameric card to have a 4% error rate.
9. James L. O'Brian, "Some Promising Approaches to Computerizing Administrative Operations," in *Eighteenth Annual Technical Conference Transactions* (Buffalo: American Society for Quality Control, May 4-6, 1964).
10. Eugene Kirby, "Quality Control in Banking," in *Administrative Application Division of the American Society for Quality Control 1975 Yearbook*. (Milwaukee, Wisconsin, 1975), pp. 55-57.
11. *Internal Revenue Code*, Section 6001.
12. Especially the "ph" series of the ANSI standards obtainable from their New York office and standard MS-110 obtainable from the National Micrographics Association, Silver Springs, Maryland.
13. *Control Procedures in Microfilm Processing*. (Rochester, New York: Eastman Kodak Company, 1974).
14. *Military Standard 414*, "Sampling Procedures and Tables for Inspection by Variables for Percent Defective" (Washington, D.C.: U.S. Printing Office, 1957).
15. *Military Standard 105D*, "Sampling Procedure and Tables for Inspection by Attributes " (Washington, D.C.: U.S. Government Printing Office, 1963).
16. William J. Latzko, "Clerical Process Capability," in *Twenty-Fifth Annual Conference Proceedings* (New Brunswick: Metropolitan Section, American Society for Quality Control, 1973), p. 132.

3

Measuring MICR Quality

With the ever increasing volume of checks to be processed, the banking industry found that the manual methods of handling checks were no longer adequate. A committee of bankers, bank stationery printers, business machine manufacturers and other interested parties was convened by the American Bankers Association in the mid 1950s. This committee proposed the mechanization of the checks processing operation by the use of Magnetic Ink Character Recognition (MICR) [1]. This recommendation was adopted in 1956. Today all checks contain magnetic ink characters and all but the very smallest banks use these characters to process their checks in both their demand deposit and transit operations.

Although almost all MICR characters printed on checks can be read by automatic equipment, a small percentage of the characters are defective and cannot be read. When this occurs, the reading equipment "rejects" the document placing it into a special pocket for subsequent correction. Such correction can be quite costly [2].

The technique of statistical quality control has been found to be very useful in minimizing this cost. The methods used are based on an understanding of the MICR technology and the application of statisical quality control to MICR.

MICR Technology

An understanding of how MICR functions in the checks processing system is essential to the control of rejects. Fortunately, the topic is not overly complicated and can be acquired easily by reviewing the specifications as outlined in the ABA publication 147R3 [3] and its supplement [4] as well as the publications of the American National Standards Institute (ANSI) X3.2 and X9.13 [5]. The ANSI standards, a formalization of the ABA 147R3, are the official specifications used. With the exception of the vertical alignment specification (ANSI allows ± 0.014 inch displacement instead of ABA's ± 0.007 inch) and reserving position 44, the standards are the same. MICR technology is a composite of the specifications and the use of MICR characters in automated equipment such as reader/sorters.

Specifications

With the exception of the vertical alignment, the character specifications have remained unchanged since their adoption in 1956. At that time, a character set called "E-13B" was chosen to be the basic set of characters to use on checks. This set consists of 14 characters: the ten numbers and four symbols. The reason for calling this set "E-13B" is of some interest in understanding MICR. The "E" refers to the fact that it is the fifth and final set of characters considered by the committee; the "13" refers to the basic character width (0.013 inch); the "B" to the second revision which was finally adopted.

With hindsight it is easy to see that the committee would have been well advised to have waited to issue the specifications. Subsequent to the adoption of the E-13B character set another, a much superior version called "CMC-7" was developed and adopted in Europe, Latin America and many other countries. The CMC-7 set contains the full alphabet, numbers and symbols. A much improved MICR line is possible with this set. Furthermore, the recognition method is very different, being similar to the Universal Product Code used in groceries. This difference reduces the reject percentage to about 1/10th that of E-13B.

This lesson is not lost on the ANSI Committee X9, Financial Services. As reported by the committee chairman, Mr. Donald R. Monks, this committee is faced with a similar situation concerning standards for post-encoding the checks for better handling of sorts, rejects and return items among others. Faced with competing current bar code technology and an experimental magnetic stripe technology, the committee is trying to balance setting immediate standards with the desire of obtaining the best method for the next few decades [6].

It is of interest that the ANSI specifications are written in two distinct parts:

1. The physical specification of MICR characters
2. Their location on document

The physical characteristics include character formation, ink application and magnetic strength. The positional characteristics cover format, spacing and skew [7].

Character formation—This area of the specification covers the dimension and edge characteristics of the characters as well as voids, the absence of ink when it should be present. These characteristics deal with the shape of the characters and their allowable tolerances.

Ink application—Ink must be applied uniformly when printing MICR or rejects are sure to occur. This portion of the specification deals with the tolerances allowed a printer. The tolerances are very minute. It must also be realized that the sensing of the character is through its magnetic feature. Any extraneous ink on either the *front or the back* of a MICR document is detected by the sensing head in the machine and will cause the item to be rejected. (See Figure 3.1.) Normal standards of printing are not acceptable when making MICR documents. Paper also plays a significant role: the surface characteristics of the paper can cause poor MICR printing to occur.

Magnetic strength—This is the main feature of MICR. It is also the feature that makes printing the checks so difficult. Magnetic sensing was chosen over optical sensing because the reader has the ability to correctly detect characters in spite of signatures, background or similar obstructions which would affect optical reading equipment.

In order for the reader to sense the magnetic character, the signal level or magnetic intensity of each character must be the correct value. To some extent this is a function of the ink used. It is also a function of the printing press operation. If the characters are printed by the

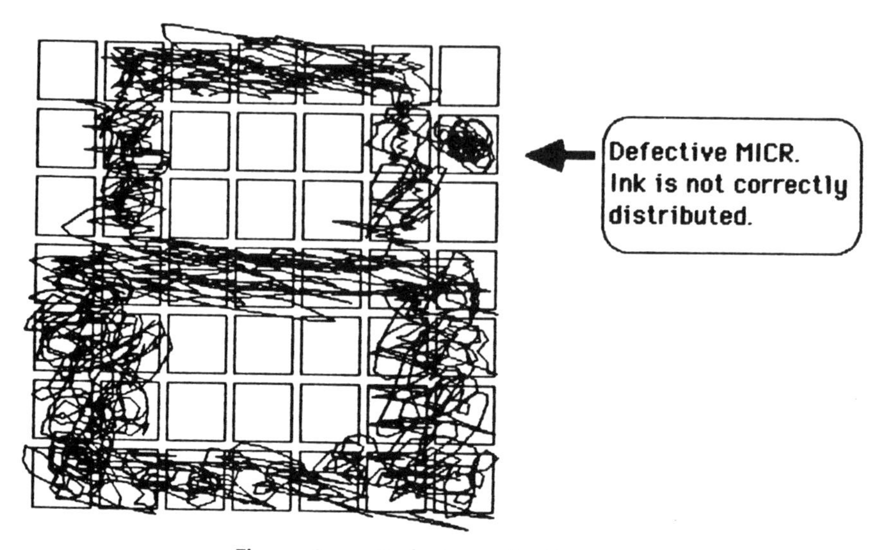

Figure 3.1 Defective MICR character.

encoding or letter press method described below, it is possible for careless presswork to cause them to be "embossed" into the paper. By embossment is meant that the character hit the paper so hard that the fibers were crushed and the ink driven below the surface of the paper. When the ink lies as little as 0.001 inch below the surface of the paper, the specifications are not met and the document is likely to reject.

Format—There are five primary fields on each check. (See Figure 3.2.) In the order in which they are read: the amount field, the on-us field, the transit field, the external processing code field and the auxiliary on-us field. The on-us and auxiliary on-us field are used at the discretion of the bank. The on-us field is often broken into three parts: process control, personal check serial number and account number. Each primary field has a specified location within small tolerances. This allows all sorters (as the reader/sorter is often called) to search for essential information in known locations. Without such a feature, the transit, or check exchange, operation between banks would not be possible.

The bottom 5/8th inch of the document is called the "clear band." It is reserved for the MICR only. No magnetic ink printing, other than

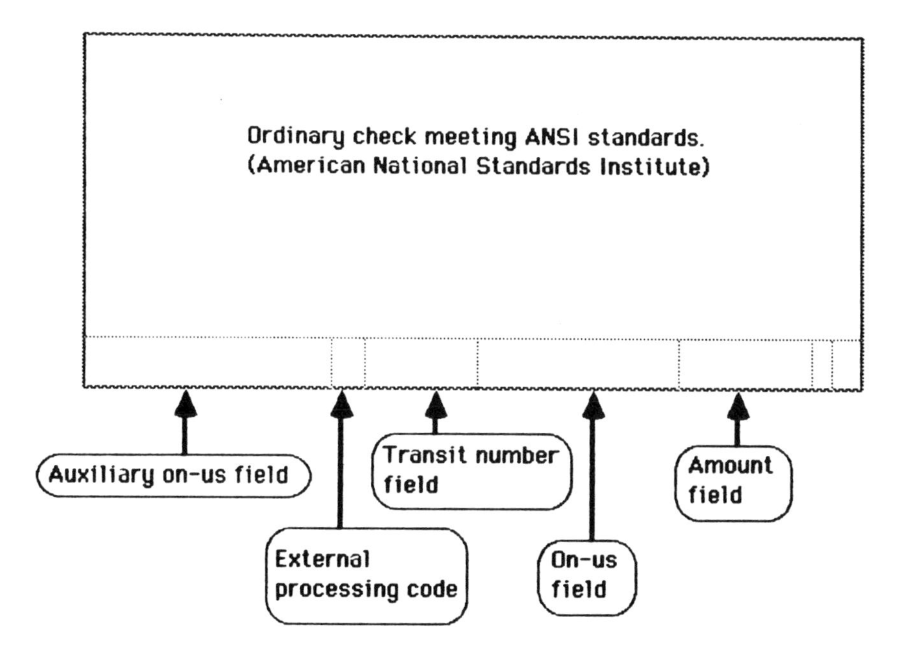

Figure 3.2 Layout of five primary fields on a check.

the appropriate characters in their correct location is permitted in this area. Sometimes, printers are not aware of this restriction and print cutting or perforation marks in the clear band. This will cause rejects.

Since all referencing of location is from the right edge and bottom of a check, it is important that the printer trims the checks properly. An otherwise excellent print job can be utterly destroyed by a defective cutting job in the bindery.

Spacing and skew—Perfect characters placed in an imperfect manner on the check can cause it to reject. The symbols and numbers must be properly positioned in relation to one another as well as in their proper location using the reference edges on the check. Horizontal and vertical alignment must be maintained within their narrow limits. The individual characters must be straight in relation to the bottom edge (skew). With all of these requirements it is a tribute to the art of bank stationery printers that they can maintain such good quality.

The Reader/Sorter

The machines which can sense the MICR and transmit this information to the computer are generally known as "reader/sorters." This electronic gear not only reads and analyses MICR to transmit the data but can also be used as stand alone units to sort the checks using the encoded MICR.

One source of puzzlement found with these machines is the apparent inconsistency with which rejects occur. If a set of documents which contain a small amount of rejects is processed through a reader/sorter a number of times, it can be noted that the number of rejects differ each time and that the items rejected are not necessarily the same each time. In fact, some software reject control methods such as CPCS (IBM's Check Processing Control System) and IPS (Burrough's Item Processing System) capitalize on the fact that, on average, 50% or more of the rejected checks can be successfully read when processed a second time.

The reason for this apparent anomaly lies in the fact that the reader/sorters are capable of reading MICR which is slightly defective. For instance, an embossed character causes the item to reject because the ink cannot get sufficient magnetic strength since it is recessed into the paper. The gap between the check transport and the read head has some dimensional tolerances to allow for different thicknesses of paper as well as the use of "carrier" envelopes (to contain torn or other bad documents). Because of this tolerance, documents may pass close to or further away from the magnet and the sensing (read) head. If they pass close enough, there may be enough magnetic flux to allow the sense head to properly decode the character; otherwise, the document rejects. Since the chance of passing close enough is about 50%, it is not surprising that 50% of once-rejected items read the second time through. (See Figure 3.3.)

Perfect characters (i.e., meeting the specifications) will read every time on a properly adjusted sorter. Utterly defective characters (e.g., MICR characters printed in non-magnetic ink) will always reject. The readability of characters falling somewhere between perfection and utter failure depends on both the degree of imperfection and the method of entering the read head system.

The ABA in its supplement to 147R3 warns against the use of reader/sorters as the sole source for testing checks [8]. In addition to

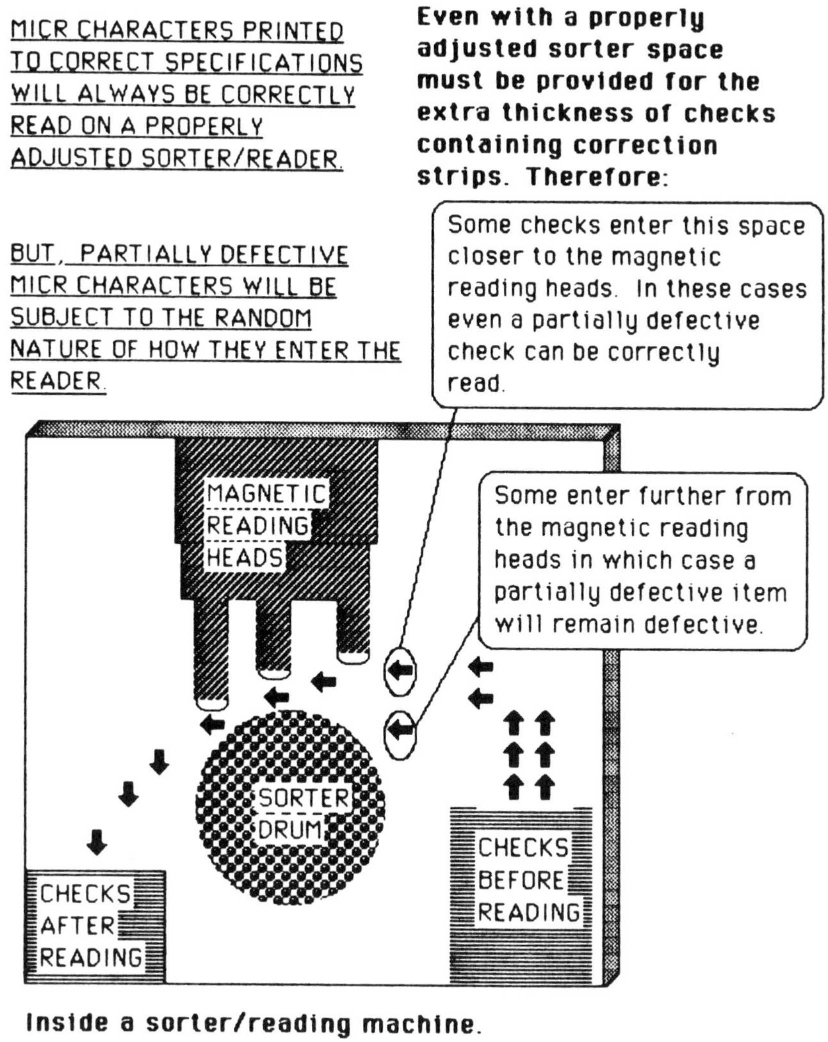

MICR CHARACTERS PRINTED TO CORRECT SPECIFICATIONS WILL ALWAYS BE CORRECTLY READ ON A PROPERLY ADJUSTED SORTER/READER.

Even with a properly adjusted sorter space must be provided for the extra thickness of checks containing correction strips. Therefore:

BUT, PARTIALLY DEFECTIVE MICR CHARACTERS WILL BE SUBJECT TO THE RANDOM NATURE OF HOW THEY ENTER THE READER.

Some checks enter this space closer to the magnetic reading heads. In these cases even a partially defective check can be correctly read.

MAGNETIC READING HEADS

Some enter further from the magnetic reading heads in which case a partially defective item will remain defective.

SORTER DRUM

CHECKS BEFORE READING

CHECKS AFTER READING

Inside a sorter/reading machine.

Figure 3.3 Why an item rejected once may be accepted next time through.

the possibility of accepting defective items, reader/sorters have another peculiarity which makes their use for testing dangerous: most sorters do not sense MICR until their system is activated by a cueing symbol (dollar amount, on-us, or routing symbol). When testing unused checks, the dollar amount field is usually blank. The ordinary sorter will ignore any magnetic ink (such as a perforation guide mark) until a cueing symbol is encountered. Some sorters have been adapted with special hardware to search for a dollar amount symbol and if not found in the proper position to generate a phantom symbol. Such a sorter can be used as a coarse screening device. Measuring conformance to MICR standards requires a special set of tools which can be found in every important bank [9].

Another point to consider is that different make sorters are sensitive to different types of defects. For instance, sorters which use a matrix of tiny sensing devices to detect the presence of magnetic flux seem to be sensitive to embossment while sorters using the gap method of wave form analysis are more sensitive to voids and defects related to character formation. Because checks may be processed by banks and the Federal Reserve System using either type of sorter, it is necessary to test checks to ABA standards rather than to one sorter or another.

The exclusive use of sorters to test MICR quality can be misleading and is unacceptable.

The Printing Process

As can be seen from the description of MICR specifications, the printing of these characters is not a simple task: it requires the utmost of the printer's art and skill. Printers who have never encountered MICR before tend to underestimate the difficulty involved in this task. The ink alone is so different from the ordinary printing ink that optimists should be on their guard. The addition of iron, nickel, or cobalt oxide to achieve the magnetic strength gives the ink undesirable characteristics. It is more viscous and does not cover well. Printers who are unaware of the ink's property try to make the ink more useful by the addition of thinners and driers. As little as 3% solvent and/or driers added to the ink can cause the critical loss of magnetic flux [10].

Added to these complications are the close tolerances which are required. There are several ways to produce MICR documents: Letterpress, encoding and offset. Banks which produce their own checks often use encoding methods. Most banks and some of the bank's customers place the MICR dollar amount on the checks presented for payment. These banks use the encoding method. The encoding method is a special case of letterpress.

Letterpress

This method of printing uses a line of type which could be set by hand but most often is made by a Linotype or similar machine. These machines use a typewriter-like keyboard to select a series of matrices which, when aligned, form a mold. Hot lead is cast into the matrices to form a slug which in turn is placed into a frame called a chase. This frame is loaded in the press, inked and paper applied under pressure. The result is a printed document.

A special case of letterpress operations exits when lithographed or offset checks are serially numbered. The numbering machines (generally used in multiples of two or more) advance after each impression. Although attached to a lithograph press, the numbering is letterpress and subject to all the problems of letterpress, perhaps even more because of the speed.

Embossment is the particular bane of letterpress operations whether printing personal checks or numbering continuous forms on a lithograph press.

Encoding

Encoding is strictly speaking a letterpress operation. The essential difference is the way in which the magnetic ink is applied to the document. While letterpress applies the ink directly to the type, the encoding process uses a ribbon in much the same way as a typewriter. The ribbon is placed between the type and the paper, then brought into sharp contact. In some machines, the paper is pushed against the ribbon and the type. In other applications, the type hits against the ribbon onto the paper.

Ribbon encoding has all of the hazards of letterpress with the additional problems created by the ribbon. Ribbons are composed of a substratum coated with a material similar to magnetic printing ink. To hold the material on the substratum requires formulation of binders that are pliable and strong. Aging tends to deteriorate some of the materials causing flaking of the "ink" which becomes extraneous ink on the document, which in turn causes rejects.

The substratum can be a problem as well. Some materials stretch rather than advance in the machine causing massive amounts of voids and character defects. In all cases, embossment and format problems are hazards to be wary of in this form of printing. Since most banks use encoding to place the dollar amount on the check, a quality control test such as outlined below is recommended. In any case, ribbon quality should be considered on the basis of performance and not on the price tag alone. (See Figure 3.4.)

Lithography

Lithography is an offset process which, through the adroit use of water and oily inks, transfers the ink image to a "blanket," a rubber sheet stretched around a drum. The image is then transferred (or offset) from the blanket to the paper.

Embossment presents no problem in this method. However, ink coverage and placement does come into play here. "The ink films are thinner, because the ink is split twice, first at the blanket and secondly at the paper" [11]. In addition the process depends on the proper mix of water and oiliness of the ink. If the formulation of either slips, i.e., the water not acidic enough or the ink not oily, the result is defective MICR.

Generally, lithography gives the most consistent good results. Unfortunately, it requires a sizable check order to make the process worthwhile. Lithography is most often used for large check orders in the thousands or greater range.

Sampling Methods

Regardless of the method of printing, there is always some risk that a batch of work is defective. As described above, the requirements of MICR printing are difficult to meet satisfactorily. This does not mean it is impossible: printers are producing excellent work every day, work which meets ABA MICR standards. But even the best printer can experience a malfunctioning of his equipment, or the inattention of a pressman. The bank must protect itself against such a condition.

If someone in the purchasing department sees ribbons for the encoding machine ON SALE, should they buy?

In actual practice, bargain encoding ribbons caused a sudden decrease in encoding productivity. Luckily, the quality control department was vigiliant and the cause of the sudden decrease in productivity was found before too much damage occured.

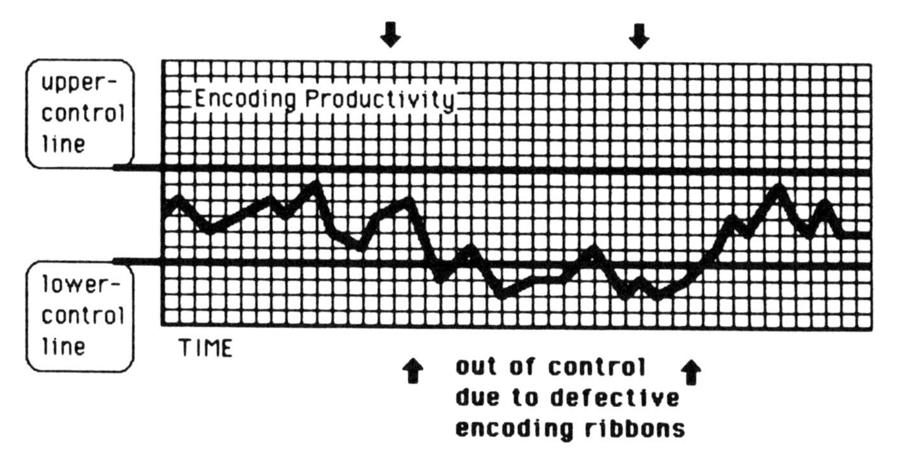

Figure 3.4 Don't purchase on price alone.

Futhermore, a number of the bank's customers have large amounts of checks made by printers, some of whom are not experienced in MICR printing, which are not known to the bank. This type of inexperienced printer frequently produces defective work to the detriment of the bank and the bank's customer.

Since the bank's customer generally has no expertise in the field of

MICR, the bank must supply this through its own expertise or through a service. The best protection for all concerned, the printer, the bank and the bank's customer, is to sample the work of the printer *prior to use*.

While there are many sampling plans available, the one developed by the government, MIL-STD-105D is satisfactory for this purpose [12]. This plan is most useful when a substantial number of documents are ordered and these can be sampled at random. The plan cannot be used or must be modified when the number of checks is small or when random sampling is not feasible. Because of these restraints there are three classes of sampling methods used exemplified by the type of document involved: personal checks, continuous form and other documents.

Personal Checks and Deposit Tickets

While this class of item could be selected on a random basis, the order quantity is generally so low (less than 500) that the appropriate sample size represents too large a proportion of the individual lot sizes. The usual personal check order is for 200 checks. The sample size required by the plan, 32, is uneconomic. There are alternatives.

It is generally not necessary to take a large sample of this type of document. Personal checks and deposit tickets are normally purchased by the bank from a selected number of vendors which can be carefully controlled. Because these printers are generally bank stationery printers, they are well versed in the printing of MICR. They almost always have a substantial quality control program of their own since they know how difficult it is to maintain adequate control over the quality of MICR printing. If they can furnish control charts of their process, that is all that the bank needs for their protection. If not, a small number of samples for use in monitoring the process of the printer, is all that is needed.

For personal checks and deposit tickets, the printer is asked to supply a single item of each order. The debit ticket used to charge the customer's account for the check printing or a sample check as well as a single deposit ticket is used as the sample. It is important that all the fields of the MICR line be represented on the sample. Since the printer has a proven track record of quality, the orders are not held pending analysis but are shipped at once. This prevents delay in servicing the bank's customer.

The documents are tested in groups and only major defects such as wrong account number are reprinted. In the case of reprints, it is up to the account officer to retrieve the defective items and replace them with good ones.

The printer's quality level is recorded continuously in the form of a process control chart. These charts tend to average around 0.25% [13]. Should a printer's chart signal an out-of-control condition, the bank can inform him at once to examine his process.

Continuous Forms

Continuous forms present a unique sampling problem. The lot quantity is usually quite adequate for economical sampling. However, the fact that the form is continuous and is also serially numbered, prevents the use of random sampling. Extracting random samples would leave unacceptable gaps in the form where the randomly selected checks were removed.

To accommodate this situation and still protect both the bank and the printer, a special sampling plan has been devised which takes into account some of the characteristics of the printing process used to create continuous forms.

Continuous forms are normally produced on lithography presses using a web — continuous roll of paper — method. A characteristic of this form of printing is that it is either very good or very bad. Once a press is set-up it continues to run steadily, good or bad. Occasionally something happens in the run to cause a problem. Generally this problem persists until the press is stopped and corrected. It is rare that random defects occur during the run.

Because of this, an examination of a sample of the work from the beginning and end of each normal press stoppage (i.e., for a paper change) will give a clear picture of the quality of the run. A press has to get up to speed anyway and the items produced during such a start-up are discarded under control. A portion of these remainders, as they are called, can be used for the sample.

How much sample should be taken? Four impositions from the beginning and end of each natural press stoppage. An imposition is the amount printed in a full rotation of the press cylinder. Generally, the printing plate will contain a large number of images of the check on one plate. Numbering machines are usually coordinated with the number of images on the plate. Four impositions will give a good picture of the quality at each end of the stoppage.

Other Items

A large number of checks and other MICR documents are prepared as individual units. These are either single or multi-part forms. The test documents are voided and, therefore, the test is considered destructive. In the case of multi-part forms with MICR on the second or later copy, the forms must be disassembled and, again destroyed. When the testing is destructive, the theory of "All or Nothing" sampling does not apply [14]. Documents of this type lend themselves to control by acceptance sampling using the standard plans available.

A commonly used sampling plan is MIL-STD-105D, "Sampling Procedure and Tables for Inspection by Attributes" [12], a plan developed by the government and used in many countries of the world. While there are better plans available, the ease of obtaining MIL-STD-105D and the common usage of the plan in this country makes it useful. Vendors should be cautioned that they carefully read MIL-STD-105D so as not to misuse it. In case of doubt they are well advised to get competent, professional help from a qualified statistician.

It is of course possible that a printer has a good quality control system in place. In such a case, the printer's control charts sent with a specimen of the check will serve the bank better than any sampling plan. Again, competent professional judgement is needed to determine whether sampling is required.

When acceptance sampling is indicated, arrangements are made with the printer to supply the appropriate number of items sampled in accordance with the plan and submit them for MICR testing. Usually it is relatively easy for the printer to select the samples at random before they are packaged. The samples are tested as described below and the lot accepted or rejected on the basis of the test results.

It is in the printer's best interest to furnish as representative a sample as possible so that a fair evaluation of the lot can be made. In the case of a large lot, it is often to the printer's benefit to split the order into logical sub-lots, such as a sub-lot of each press run, for each shift and each press. This will avoid rejection of the whole lot due to a defect in an isolated portion of the order.

Pre-Delivery Sampling

A number of banks, in purchasing their large run checks and other MICR documents, require the printer to select a sample for testing as part of their purchase order requirements. If the use of MIL-STD-105D was specified together with the information on level, single sampling (or other) and AQL, the printer can select the sample as the work is produced and send it for approval before the lot is shipped. Should it be necessary to reprint, the shipping expenses have been saved as well as the pressure to accept a defective lot, a pressure often exerted when the defective items are already in the warehouse.

Based on analysis of many past orders it has been found that,

> . . . using Level I, AQL = 0.4%, Single Sampling will cover most needs from an economic point of view. Printers experience no difficulty in meeting these standards while the long run protection to the bank is a maximum of 1% (AOQL) defectives for fresh products. The general quality using this plan in practice was well below the 1% level. This in spite of the fact that the reject rate is increased by the handling the checks received from the general public and some banks in the system [13].

Documents sent in for pre-delivery sampling (PDS) are tested by pre-screening on a sorter adapted to detect all MICR. Defective items and samples of all items are further analyzed. This analysis requires some pieces of relatively sophisticated apparatus to determine conformance or non-conformance to ABA standards. These are outlined in a number of articles [15].

The fact that a sample of a lot was accepted does not relieve the printer from responsibility for the printed lot as a whole.It is always possible that the sample was not selected properly and so not representative of the lot. It may even have come from a different lot. But even if the selection was proper, the vendor is still responsible for his product.

Control of Operations

Acceptance sampling is used to guarantee good raw materials for use in the processing of checks. It does not help in maintaining the quality of the operation. To control the quality of the process it is necessary to use process control techniques.

Process Controls

As described starting on page 25 the control of a process is best accomplished with a Shewhart Control Chart. In the case of check rejects, the most obvious form of control consists of a p-chart of the rejects observed at the reader/sorter. (See Figure 3.5.) The p-chart, or fraction defective chart is a graphic record of the number rejected divided by the total processed. The upper control limit is computed by adding 3 standard deviations (3s) to the overall average, P. (UCL = P + 3s). The standard deviation is computed with the formula:

$$s = \text{standard deviation} = \sqrt{P^*(1 - P)/n}$$

or in words

s = standard deviation = square root of (P times (1 − P) divided by n)

where n = total number processed on any given day.

The lower control limit is computed in either of two ways depending on whether the 3 standard deviations (3s) are greater or equal to the average, P. If the 3s are greater or equal to P, the lower control limit is zero (0). If the 3 standard deviations are less than P, the lower control limit is found by subtracting the 3 standard deviations from P. (LCL = P − 3s).

As pointed out by Cowden among others, just because it is possible to form percentages does not mean that the underlying assumption of a binomial distribution is true [16]. The nature of check processing is such that the underlying assumptions for the proper use of the p-chart are not justified: the value of n is very large and is really a mixture of many accounts pulled together. To compensate for the problem the X-bar and R-charts are used instead of the p-chart.

To use the X-bar, R-charts the daily reject rates are recorded for the week. Each week is considered as a sample of five (sometimes four) daily observations. The control limits are computed in the normal manner described in numerous texts on quality control [17].

Day	Obs.	Sample	# Def.	P	LCL	Avg	UCL
Monday	1	930	8	0.86%	0.06%	1.00%	1.94%
Tuesday	2	1080	12	1.11%	0.06%	1.00%	1.94%
Wednesday	3	1050	16	1.52%	0.06%	1.00%	1.94%
Thursday	4	1020	4	0.39%	0.06%	1.00%	1.94%
Friday	5	1050	8	0.76%	0.06%	1.00%	1.94%
Monday	6	1040	14	1.35%	0.06%	1.00%	1.94%
Tuesday	7	920	11	1.20%	0.06%	1.00%	1.94%
Wednesday	8	1000	16	1.60%	0.06%	1.00%	1.94%
Thursday	9	990	4	0.40%	0.06%	1.00%	1.94%
Friday	10	950	8	0.84%	0.06%	1.00%	1.94%
Monday	11	970	13	1.34%	0.06%	1.00%	1.94%
Tuesday	12	950	4	0.42%	0.06%	1.00%	1.94%
Wednesday	13	1030	6	0.58%	0.06%	1.00%	1.94%
Thursday	14	980	13	1.33%	0.06%	1.00%	1.94%
Friday	15	1050	4	0.38%	0.06%	1.00%	1.94%
Monday	16	1070	12	1.12%	0.06%	1.00%	1.94%
Tuesday	17	980	11	1.12%	0.06%	1.00%	1.94%
Wednesday	18	940	19	2.02%	0.06%	1.00%	1.94%
Thursday	19	1050	11	1.05%	0.06%	1.00%	1.94%
Friday	20	950	6	0.63%	0.06%	1.00%	1.94%
	Total =	20000	200	1.00%			
	Average=	1000					

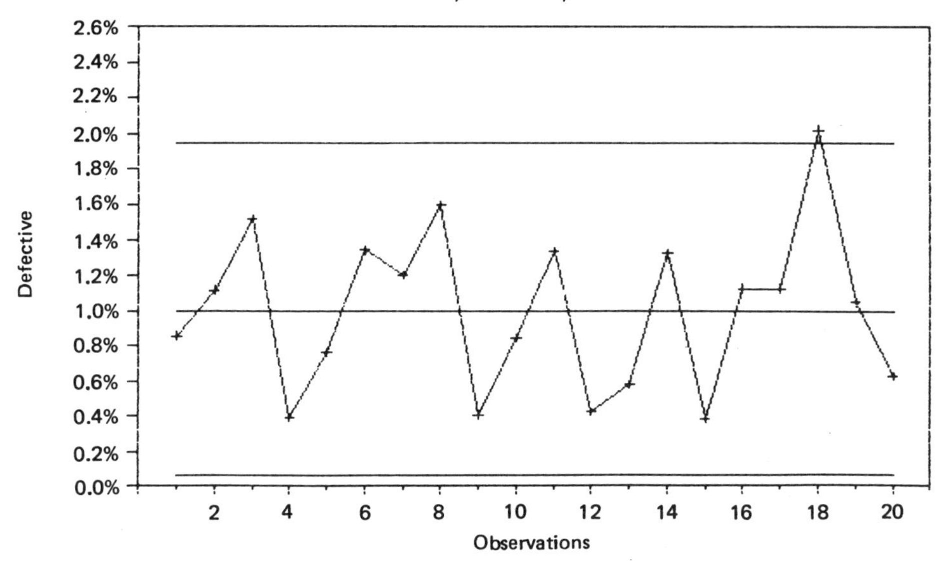

Figure 3.5 Example of p-chart data.

In addition to control charts, process control theory requires that equipment be capable of performing the task assigned to it. To determine this, a special method called set-up check is used. An application of this method to MICR quality control is described in the next section.

Since the machines are operated by people who can contribute to the reject rate, it is important to use process controls which can separate the source of rejects, the system or the operators. Generally it is the system that causes the problem; operator problems are a tiny amount of the total. This topic is treated fully in Chapters 4 and 5.

Encoder Quality Control

Encoders (or as they are called by IBM, inscribers) are machines which can place MICR numbers and symbols on checks and other paper documents. These machines range from relatively simple, small devices to sophisticated, computer driven models with built-in logic capable of handling thousands of documents per hour. Normally, encoders are used to place the dollar amount on a check. However for internal tickets, repair work, starter sets and the like, encoders are called upon to provide a number of fields up to a full line.

The basic concept of control is the same whether the encoder is a manual device or a sophisticated machine. Samples of the encoded work are taken at set intervals and examined. If the samples show encoding defects, the machine is taken out of service until a mechanic has repaired it, as evidenced by further samples.

Manual encoders are usually found in small banks and branch operation. They present a special problem. Their normal use is to prepare starter kits, small batches of checks or deposit tickets, and intermittent other MICR requirements. They are usually used by personnel that have only been casually trained, if at all, in the use of the machine. In fact, the machine appears to be so simple to operate that untrained personnel have been known to use it. The selection of MICR characters on such machines is normally by moving levers to appropriate positions, inserting the item to be encoded and either pressing a button or hitting a sensing switch to activate the system.

The four most common problems found in such machines are

1. Wrong placement of MICR
2. Wrong use (or absence) of symbols
3. Embossment
4. Extraneous ink from ribbon flaking

Many banks have removed these devices from their branches and use central encoding units because they found that the quality was too hard to maintain in the branches.

Those banks that still use these machines in their branches have found it useful to take samples of the work produced at frequent intervals, as often as every day, and to test these sample documents. Another method of control commonly used is to obtain a sample of every starter kit made on these machines and to test the samples before they are permitted to be used. This method prevents many "bad account number" failures. A further precaution is to rotate the ribbon stock. Ribbons should not be older than six months for optimal quality.

The automatic encoders present other problems. As high speed production tools, they are susceptible to normal machine wear. It is important that a record be kept of the running time of each machine and a log of the preventative maintenance. It is important that the customer engineer (CE) adhere to the manufacturer's schedule of maintenance to achieve proper quality levels.

Apart from such elementary controls, each encoder should be tested at the start of each shift by passing a number of encoder test documents (see Figure 3.6) through the machine. The number of documents to be used depends upon the type of machine: ten key machines need three to five documents with each character represented; multi-key machines need ten documents, one for each number. This test serves two purposes:

1. To check the quality of the encoding
2. To clean the hammers of any extraneous ink that might have accumulated due to the failure of a document to pass through the transport

For the same reason it is a good idea to perform a start-up test whenever a ribbon is changed on the machine.

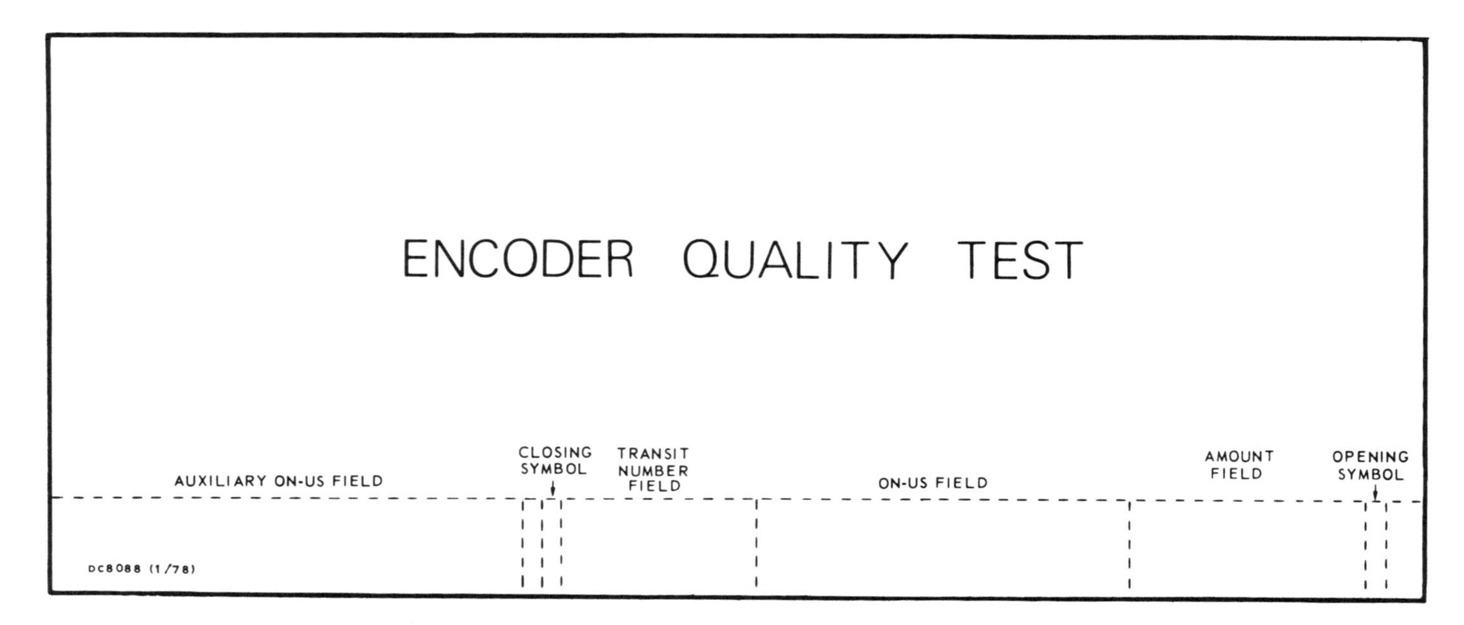

Figure 3.6 Encoder test document. Reproduced here at 57% of original size.

The test documents must be examined as soon as they are produced. This can be done by the shift supervisor. A visual examination of the MICR line is useful: if voids, extraneous ink or embossment can be seen by the naked eye, they probably exceed the ABA standards. An Edmunds magnifier can be used to confirm suspected deviations from standard. A layout or position gauge can be used to check the proper placement of MICR fields [18]. These pieces of equipment are inexpensive enough to be available in every operations center. It is strongly recommended that all supervisory personnel connected with checks processing be furnished these tools.

If the test documents indicate the presence of defective MICR, the machine should be taken out of service until repaired by a mechanic. It is recommended that the test documents be sent to a MICR test laboratory, either within the bank or to an outside service, to test for problems that are not readily detected visually such as magnetic intensity, embossment, tiny character defects among others [19].

It is advisable to maintain a record of the number of checks processed between failures of the machine. Except for random fluctuations, the machine should have the same process rate between failures. A lower than expected process rate is a signal of some major problem either with the machine or the materials processed. This should be brought to the attention of the mechanic servicing the equipment. If he cannot solve the problem, the manufacturer's engineering staff should be consulted. Experience shows that a persistent problem signalled by the above mentioned control chart frequently requires major engineering effort to overcome [20].

Sorter Control

The automatic input machines which read the MICR documents are also highly sophisticated mechanical devices. They are subject to wear in the same manner that encoders wear out. Therefore, it is important to use similar controls on the reader/sorters.

The first line of defense is appropriate preventative maintenance. Strange as it may seem, preventative maintenance can be overdone. The proper way to do preventative maintenance is to leave the machine alone until a control chart of measured characteristics signals an out of control condition. Then the correction should be made. The alternative use of maintenance after so many hours of operation is not only wasteful but also hazardous. *Tinkering with a system that is in control can cause it to go out of control.*

A set-up check for the reader/sorter is recommended. The operator is responsible for the cleaning of the machine and setting the switches correctly. A set-up test consisting of a test deck of documents with known characteristics can disclose potential problems in the setting of the machine. If these occur, the CE should be notified at once to correct the indicated problem.

A continuous monitoring of the reject rate was described above. This is highly recommended. However, the monitoring is useless if no action is taken when special causes are signalled. The effectiveness of a control chart depends entirely on its proper use. Prompt analysis is essential to the proper application of a Shewhart chart.

References

1. *The Common Machine Language for Mechanized Check Handling*, Publication 147R3, American Bankers Association, New York, 1967.
2. Latzko, William; Leszczynski, Walter; Morogiello, Anthony; O'Leary, Jack; and Sullivan, Kenneth. *MICR Quality Control Handbook*. Washington, D.C.: American Bankers Association, 1982.
3. *The Common Machine Language for Mechanized Check Handling*, Publication 147R3, American Bankers Association, New York, 1967.
4. *Supplement to the Common Machine Language for Mechanized Check Handling* (147R3), American Bankers Association, Washington, D.C., 1971.
5. *Print Specification for Magnetic Ink Character Recognition*, X3.2-1970 and *Specification for Placement and Location of MICR Printing*, X9.13-1983, American National Standards Institute, New York. Specification X3.2 is in process of revision by Committee X9, Financial Services.
6. From a presentation by Donald R. Monks, EVP Irving Trust and chairman of ANSI Committee X9, at the Bank Administration Institute's "MICR Document Testing Workshop," Baltimore MD, December 13-15, 1984.

7. William J. Latzko, "Quality Control of MICR Input," in *Twelfth Annual Conference Proceedings* (Hempstead, NY: Long Island Section, American Society for Quality Control, 1974) pp. 38ff. This paper contains details concerning the specifications.
8. *The Common Machine Language for Mechanized Check Handling*, Publication 147R3, American Bankers Association, New York, 1967, p. 9.
9. William J. Latzko, "Quality Control of MICR Input," in *Twelfth Annual Conference Proceedings* (Hempstead, NY: Long Island Section, American Society for Quality Control, 1974) p. 41.
10. William J. Latzko, "Why Banks Need Bullseye Accuracy and Consistency," *Printing Impressions*, Vol. 20, No. 7 (December, 1977) p. 102.
11. E. A. Apps, *Ink Technology for Printers and Students*, Vol. III, Chemical Publishing, Cleveland, Ohio, 1964.
12. *Military Standard 105D*, "Sampling Procedure and Tables for Inspection by Attributes " (Washington, D.C.: U.S. G.P.O., 1963).
13. William J. Latzko, "Quality Control of MICR Input," in *Twelfth Annual Conference Proceedings* (Hempstead, NY: Long Island Section, American Society for Quality Control, 1974) pp. 42–44.
14. Deming, W. Edwards, *Quality, Productivity and Competitive Position*, Cambridge, MA: Massachusetts Institute of Technology, Center for Advanced Engineering Study, 1983, Chapter 13 and Latzko, William J.,"Minimizing the Cost of Inspection", in *Thirty-Sixth Annual Quality Congress Proceedings* (Detroit: ASQC 1982).
15. Latzko, William J., "Quality Control of MICR Input," in *Twelfth Annual Conference Proceedings* (Hempstead, NY: Long Island Section, American Society for Quality Control, 1974) p. 48 and Latzko, "MICR Challenge for Bankers," p. 8 of the text of the speech.
16. Cowden, Dudley J., *Statistical Methods in Quality Control*, (Englewood Cliffs, NJ: Prentice-Hall, Inc., 1957), pp. 373ff.
17. See for instance, *ASTM Manual on Presentation of Data and Control Chart Analysis* (4th edition, Philadelphia, PA: ASTM Special Technical Publication 15D, 1976), pp. 104-105.
18. Magnifier #330285 with E-13B Reticle from Edmund Scientific

Co. Barrington, NJ and "Position & Dimension" gauge M1029-34(R) from Clearwave Electronics, Niagara Falls, NY.

19. William J. Latzko, "Statistical Quality Control of MICR Documents," in *Thirty-First Annual Technical Conference Transactions* (Philadelphia, PA: American Society for Quality Control, 1977), p. 118.
20. William J. Latzko, "Quality Control of MICR Input," in *Twelfth Annual Conference Proceedings* (Hempstead, NY: Long Island Section, American Society for Quality Control, 1974) p. 45.

4
Clerical Processing

Most of the early work on quality control in clerical processing originated from people working in the insurance industry, mail order operation and the government. The Bureau of the Census was particularly active in this area since they recognized early that clerical error in processing the census could have a devastating result on the country. It is not surprising that their emphasis on clerical quality control came so early since they were also in the forefront of automated data processing. While manual processing allowed for some scanty checks to take place, the automatic processing pioneered by Dr. Hollarith did away with some of these safeguards. If the input is in error, the output will also be false. "Garbage in, garbage out (GIGO)" is a common expression used by programmers and systems people to denote this phenomena.

The methods used to control the quality of clerical operations runs the gamut from simplistic error controls to sophisticated, cost saving methodology. William Exton, Jr. is a proponent of "...reducing errors by the human 'components' of clerical systems" [1]. His system of EFAR (Error Factor Analysis and Reduction) is based on the assumption that,

> In general, today's clerical employee is not as diligent, careful, efficient, or dedicated as the few "old-timers" who once constituted the bulk of the workforce. Part of the problem is that the newer workers lack the skills and experience needed to perform at a desirable level [2].

One of Exton's axioms is that "Mistakes are made by people" and that these are correctable if only once it is found out how the mistakes are made, although he does allow that, "Some causative factors are environmental" [3]. His remedy is to make a skilled study of the causes of all errors and who contributes to these errors. They in turn are studied to see what makes these people make the mistakes so that corrective action can be taken.

His system does not allow for correction of the environmental factors, just the human ones. It does not allow for errors which are faults of the system. These faults tend to be the large majority of the cause for error. W. Edwards Deming estimates this type of error to constitute 85% of the causes for mistakes [4]. Unlike Exton, Deming maintains that it is important to measure the capabilities of the clerical system and to concentrate on correcting only those operators who do not exhibit performance with statistical control. As Deming states, ". . . it is demoralizing and costly to call the attention of a production-worker to a defective item when he is in a state of statistical control. The

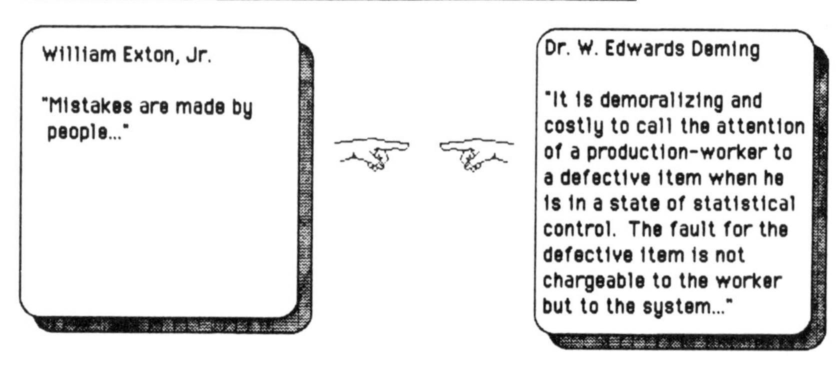

Figure 4.1 Two views of what causes errors.

fault of the defective item is not chargeable to the worker, but to the system" [5]. (See Figure 4.1.)

Most investigators have come to recognize that errors caused by the system are a major contributor to the problem and that these can best be removed by management. The workers do not have any control over the system. Old methods which failed to recognize this are being discontinued. (See Figure 4.2.)

> Over a 75-year span, many quality control techniques have been tried by Aldens. Many have been discarded; for example, the practice of levying five-cent fines against clerical workers for every error detected went out of vogue back in the Thirties. Many other practices, though still in use, have been refined and modified to meet the requirement of today's more enlightened operating and personnel policies [6].

Aldens uses a form of sampling to control the quality of their clerical staff. Most of the published methods also rely on some form of statistical quality control since this seems to be the most effective in pinpointing the cause for error and is, at the same time, most cost effective.

Many of the recorded methods of applying quality control to the clerical processes have used the technology which was so successful in manufacturing. There are, however, some substantial differences between the manufacturing and clerical forms of production.If these differences are considered, the methods used for the industrial process can be converted and modified to be perfectly suitable for clerical quality control.

Nature of the Clerical Task

The clerical task involves the "processing and output of paper" [7]. The product of a clerk is the result of some action taken such as typing a letter, entering the data into a system, completing a phone call, or advising a customer on how to handle a particular transaction. Because there are substantial differences between the processes of manufacturing and the clerical process, the quality control methods of the

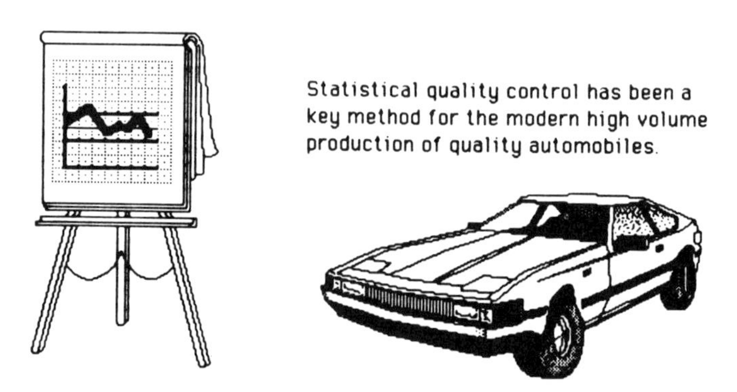

Figure 4.2 One technique from the 1930s is discontinued, another has become a worldwide standard.

former cannot be applied as directly as had been possible in the case of the high speed data processing as described in the last two chapters. This section will examine the differences, discuss clerical workflow and develop some concepts on what produces clerical quality.

Comparison to Manufacturing Processing

Clerical processing differs from manufacturing in four principal ways:

1. Specifications
2. Product
3. Measurement of output
4. State of control

By examining the nature of these differences it is possible to see how the successful techniques of technical quality control can be applied to clerical quality control. (See Figure 4.3.)

Specifications—In today's environment, no manufacturing process can take place without a blueprint specifying the exact dimensions and

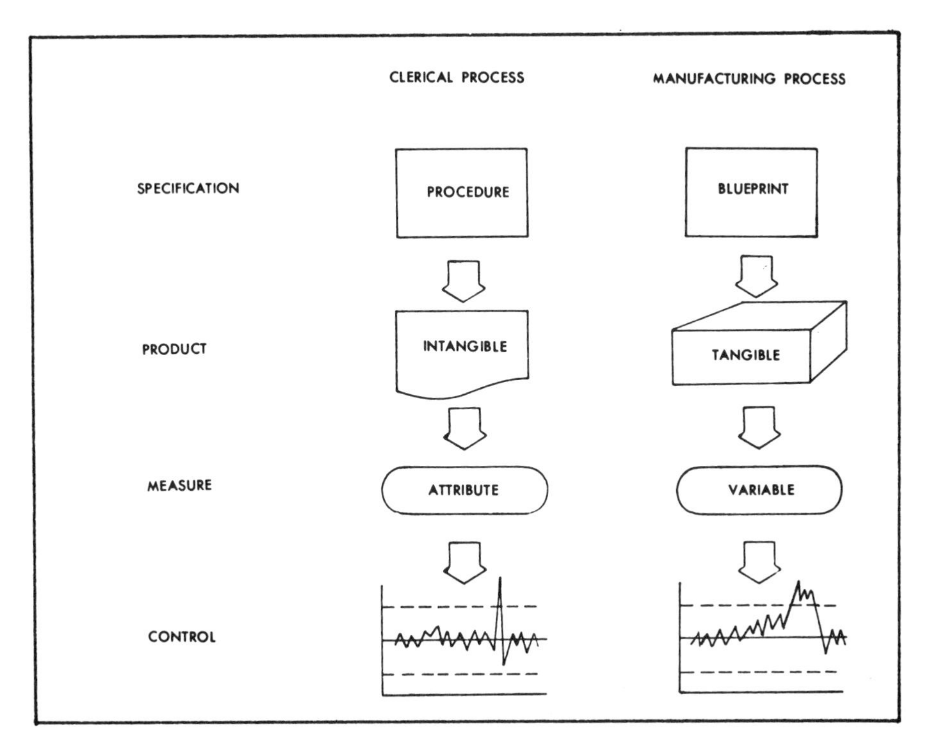

Figure 4.3 Comparison of clerical and manufacturing systems.

materials of construction. The equivalent document in the clerical process is the procedure which must be followed. Unlike the blueprint which must exist in written form, the procedure may be a verbal training received by the clerk when first starting to work. While no one would think of manufacturing an item without an updated blueprint, it is often the practice in clerical operations to perform work with old, and even without any, procedures whatever. As a result, clerks must do the best they can under adverse conditions. Many clerical errors are due to inadequate instructions such as procedures.

Product—The manufacturing process nearly always turns out a product which is tangible. Something is made that has form and is measurable in several ways. The clerical process usually turns out an intangible item, such as a decision. Compared to manufacturing "It is not quite so easy . . . to specify standards in clerical operations. For example, how does one go about determining precisely whether a dictated letter is satisfactory" [8].

Measurement of output—The product of the manufacturing operation by its nature can be measured in various physical dimensions such as length, area, capacity, mass and similar objective standards. As a result, most manufactured product can be controlled by using variable measurements, that is measurements which can be expressed on a continuous scale. The clerical process on the other hand, rarely lends itself to such a form of measurement. The best that can be done in clerical operations is to distinguish between correct and incorrect. This form of measurement is called attribute measurement.

State of control—Since much of manufacturing is mechanical, the quality of the process tends to remain in control until it gradually shifts due to tool wear and finally falls out of control. When this occurs, an adjustment brings it back into control to repeat the process. Clerical operations, however, depend on human beings. Their output tends to stay in control almost at once. There is more variability in clerical processing than in manufacturing and it tends to be more random. The analyst must adjust his method to this phenomenon.

Workflow Pipeline

The workflow concept can best be illustrated with a schematic diagram of three clerks in a department contributing to the total output of the

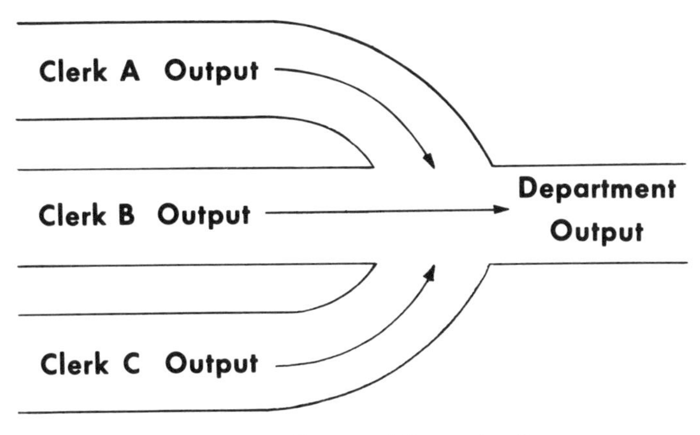

Figure 4.4 The workflow pipeline.

department. (See Figure 4.4.) It is of importance in determining the method of measuring clerical quality [9].

In the diagram the three clerks are merging their output into the department output. A large number of papers on quality control for the clerical process concentrate on the measurement of the quality of the department output. While this measure, when properly taken, gives some valuable information, the process average, it fails to give satisfactory data for the improvement of the quality of the department [9]. The process average is a coarse measure useful in tracking the performance of the department. This is part of the job of clerical quality control. To be entirely effective, it is necessary to develop a measure for the standard of quality which each clerk can achieve and is expected to perform.

The approach to this is to consider the impact of a clerk who performs at a less satisfactory level than the others. If clerk A is assumed to be worse than clerks B and C it can be readily seen that the department output will have more defective items to the extent that clerk A introduces excess defectives because of his poor work. The level at which the department can operate is really represented by the performance of clerks B and C, the process capability. The process average of the department is worse than the process capability to the extent that clerk A exceeds the process capability.

The function of clerical quality control is to first determine the process capability; second, to identify any outlier clerk such as clerk A.

A good clerical quality control system will also develop sufficient data for the operations management to locate the problem, which causes the outlying clerk to do so badly, and thereby achieve correction. This process will result in quality improvement since the department brings its process average into line with the process capability.

What Produces Clerical Quality?

Three interrelated factors have been identified: (1) training, (2) motivation and (3) management [9]. These factors are the basis for clerical quality. Others have identified the effects which result when the basic factors are not adequately provided [3].

Training—It would seem axiomatic that a clerk who is not properly trained is more likely to make errors than not. Yet, with the frequent lack of up-to-date procedures and inadequate training methods (such as letting one clerk teach the other), training is a serious problem resulting in poor quality. From a quality point of view, as well as from the production side, proper training is one of the best investments a bank can make. A Bank Administration Institute study of teller over and short shows a definite relationship in quality to training [10].

Motivation—This is a broad topic on which much has been written. It is a fact that without motivation both production and quality suffers. A very good review of this is given in the American Society for Quality Control's book *Quality Motivation Workbook* [11]. In this book, the major concepts of Mayo, Maslow, and Herzberg, among others are collated to show that quality depends on the worker having the necessary tools and management support to achieve a high level of good productivity.

Management—Management, particularly the first line manager, is the key ingredient in obtaining good quality output from the workforce. Not only are they responsible for the training and motivation of the staff but also set the standards for the operations and control them. The manager of the department sets the tone for that department. If the clerks know that high quality is expected, they will set their priorities accordingly. If no such expectation exists, there is no reason for the clerical staff to strive for quality. The expectation set must be by example and not just another exhortation for "better quality." The clerks must know that management "cares."

The Checking Process

Many managers abrogate their responsibilities for quality control to other clerks who check the work. In fact, checking is the single most important tool for quality control in most banks. Yet this significant topic has had scant attention. The assumption is that any person can do checking. This section deals with the types of checking or verification and the reliability of checkers.

Types of Verifications

There are two principal methods of checking: dependent and independent. (See Figure 4.5.) Dependent verification is more frequently encountered in clerical quality control than independent. Dependent verification is simply the comparison of the finished work to a standard such as an original order. Proofreading is a prime example of dependent verification. Independent verification is doing the work in two

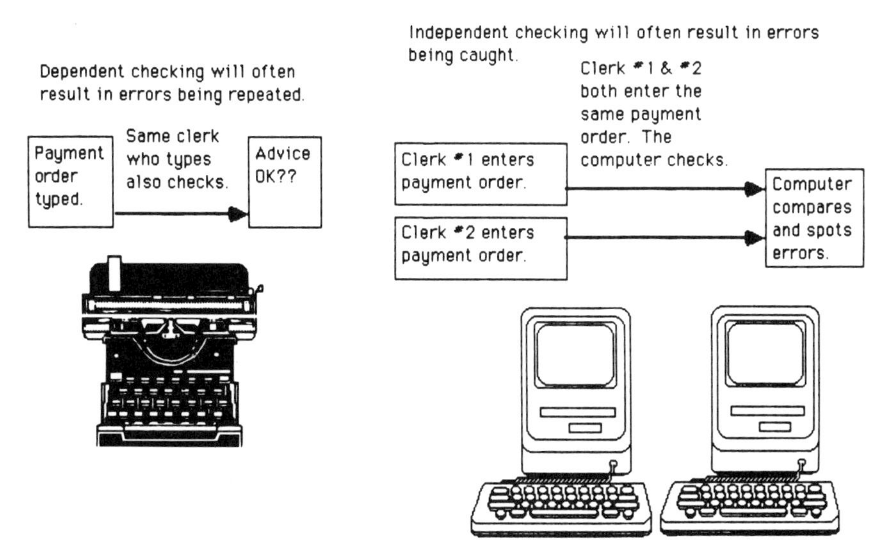

Figure 4.5 Dependent checking versus independent checking.

distinct ways and comparing the results. The use of check digits and keypunch verification are examples of independent verification.

In general, dependent verification is not as effective as independent verification. The hypnosis effect comes into play. This is the same effect that causes people to make the same mistake twice when rechecking an arithmetic operation. While independent verification is by no means a fail-safe method of checking work, it is, on the whole, much more efficient in error detection.

Dependability of Checkers

The general reliance of management on checkers to control the quality of clerical efforts raises the question of how dependable is the checking function. In order to be useful for quality control purposes, the checking function should fulfill four requirements:

1. Completeness
2. Catch all the errors
3. Return errors to the originating clerk
4. Keep good quality control records

Completeness—In general it has been found that checkers do not examine items fully. When asked to prepare a list of those things for which they check, they usually return a list which contains both omissions as well as extraneous items [12]. It has been hypothesized that this is due to the training methods employed. Generally, new checkers are trained by the incumbent checker. If during this training, error items pass before these checkers. they might well learn how to handle such errors. Otherwise, they are entirely dependent on procedures and their experience. During such training sessions the incumbent checker can pass on his idiosyncracies to the new checker. This accounts for the extraneous checking. (See Figure 4.6.)

Do checkers catch all the errors?—Studies have shown that they do not catch all the errors [12]. Their find rate depends on the type of checking and the complexity of the item checked. To illustrate the point consider the sentence shown below. Allow yourself 30 seconds to count the letter "F" which appears in the sentence.

FEDERAL FUNDS ARE THE RESULT OF YEARS
OF SCIENTIFIC STUDY COMBINED WITH
THE EXPERIENCE OF YEARS

This type of exercise is a classic example of dependent verification. In such a case it is usually better to look for the letters by scanning from the last word to the first. In that way the six letters will become apparent. The general experience with this sentence is a find

GENERALLY, NEW CHECKERS ARE TRAINED
BY THE OLD CHECKERS

"This is the way I've been doing it..."

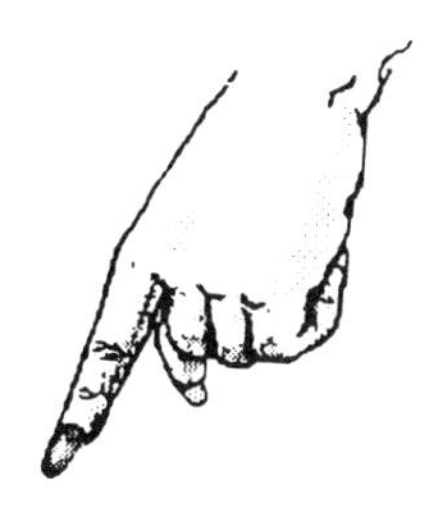

DRAWBACKS OF THIS METHOD

Old checker teaches new checker:
- incorrect procedures,
- wrong information,
- inadequate methods,
- various other idiosyncratic behaviors.

Figure 4.6 How checkers get trained.

rate of three to four letters which corresponds closely with the experience of a 60% to 70% find rate for checkers [12].

Are the errors returned to the originating clerk?—It is generally found that errors are returned to the section making the mistake. There the supervisor may or may not return these to the originating clerk. In most cases, the pressure of time is such that the item gets corrected and sent out of the shop as quickly as possible to meet deadlines. It is already delayed by virtue of the rejection. In a number of cases, checkers make their own corrections. These have a good chance to be incorrect and leave the bank without any further checking. (See Figure 4.7.)

Do checkers keep good quality control records?—In general, checkers do not maintain quality control records. In those cases where records are kept, they generally are a summary of errors found, sometimes

This flow chart shows an IDEAL SITUATION in which errors are returned so that immediate feedback is provided.

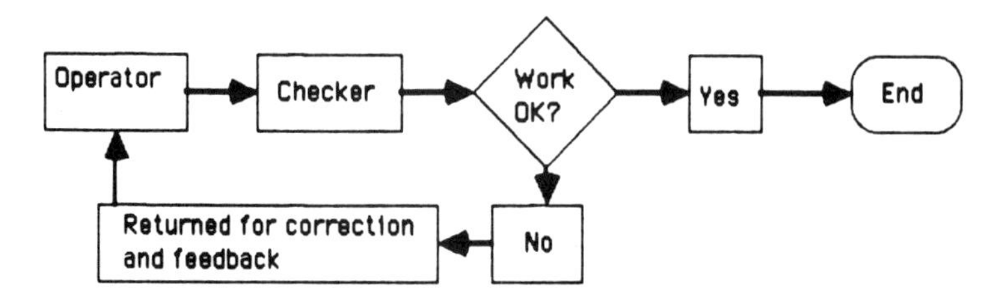

This flow chart shows what GENERALLY OCCURS. When errors are found the work is usually late so the checkers make the correction and feedback is not provided until well after the error is made. No one learns their job better, and mistakes continue to occur.

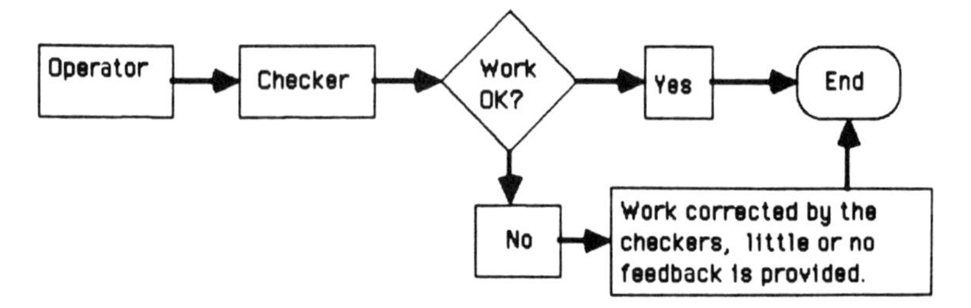

Figure 4.7 Why individual errors get fixed but not prevented.

listed by type. On the whole, these records are unsatisfactory since they do not reflect the source of errors in such a way that measurements can be made.

Record keeping, like returning defective items to a clerk, should not be a function of checking. The effect desired can be better obtained through the QUIP method described in the next chapter. Checkers should concentrate on knowing what to look for and finding errors [12].

References

1. William Exton, Jr., "How error-prone is your bank?" *Banking*, May, 1977, p. 52.
2. William Exton, Jr., "How to Improve Clerical Accuracy," *Supervisory Management*, April, 1971, p. 30.
3. William Exton, Jr., "How error-prone is your bank?" *Banking*, May, 1977, p. 56.
4. W. Edwards Deming, "On Some Statistical Aids Towards Economic Production," *Interfaces*, Volume 5, No. 4 (August, 1975) p. 3.
5. W. Edwards Deming, "On Some Statistical Aids Towards Economic Production," *Interfaces*, Volume 5, No. 4 (August, 1975) p. 1.
6. V. N. Andersen, "Five steps to quality control of clerical operations," *System & Procedures Journal*, November-December, 1964, p. 8.
7. Thomas C. Staab, "Quality Applicable to Paperwork? Probably!" in *Twenty-Seventh Annual Technical Transaction* (Milwaukee: American Society for Quality Control, 1973), p. 393.
8. Bennett B. Murdock, "Quality Control in Clerical Operations," in *Leadership in the Office* (New York: American Management Association, Inc., 1963), p. 243.
9. Latzko, William J., "QUIP—The Quality Improvement Program," in *Twenty-Ninth Annual Technical Conference Proceedings* (San Diego: American Society for Quality Control, 1975) p. 248.
10. James Bergstrom, *Teller Differences Rate. A Study of Factors*

Affecting Teller Performance, Publication #700 (Park Ridge, Illinois: Bank Administration Institute, 1976), p. 13.

11. American Society for Quality Control, *Quality Motivation Workbook* (Milwaukee, Wisconsin, 1967).
12. Latzko, William J., "QUIP—The Quality Improvement Program," in *Twenty-Ninth Annual Technical Conference Proceedings* (San Diego: American Society for Quality Control, 1975) p. 249.

5

QUIP—The Quality Improvement Program

Quality improvement depends on bringing the quality of clerical operations into control, that is to say within the process capability, and then determining the cost effectiveness of modifying the system to improve the process capability. The first step to developing optimum quality is to bring the operation into a state of control. This is achieved by eliminating the special causes, such as untrained operators through training, and maintaining the stability of the system by constant checking. The single most important person in this process is the first line supervisor.

The Supervisor's Role in Clerical Quality

The person most involved with the training and motivation of the clerks in a department is their first line supervisor. This person is the key to clerical quality. If the supervisor fails to take an active part in the processing of the work, constantly monitoring and controlling the operation, there is little incentive for his employees to be concerned with the quality of the output. To develop the role of the supervisor it is necessary to examine his responsibilities and the actions he should take in achieving quality.

Responsibility

The supervisor is responsible for the quality of his operation. It is true that he is also responsible, among other things, for production, personnel matters and planning. But he *is* responsible for quality and the other responsibilities should not be put forth as an excuse to evade any one of them (see Figure 5.1).

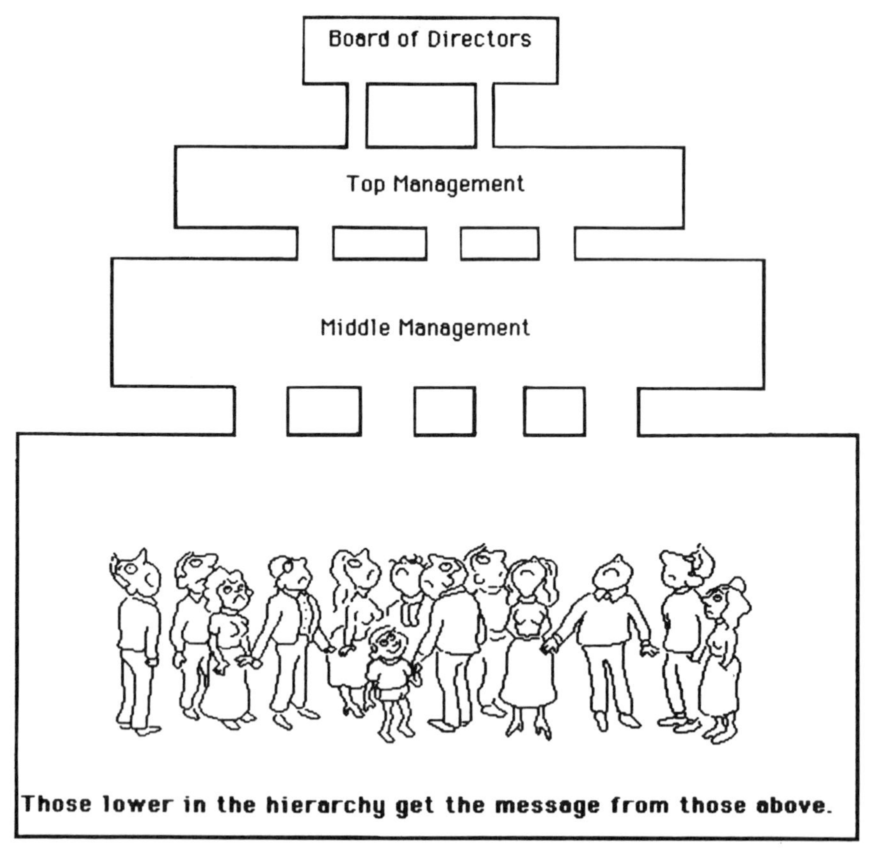

Figure 5.1 Quality is the responsibility of everyone.

The difficulty is that many first line supervisors get on-the-job training and do not really know how to handle some of their more complex problems such as time. Work as hard as they will (mostly on production) they never seem to have enough time (see Figure 5.2). It is obvious to them that their primary function is productivity, but it is the rare supervisor who recognizes that this means 'good' production. The beginning supervisor does not usually see the connection between quality and costs such as overtime. It seems that by the time they come to recognize these factors and learn to handle time, they are promoted and another beginning supervisor takes their place.

Responsibility for quality is shared by everyone in the organization from the clerk to the first line supervisor through all levels of supervision and management to the chairman of the bank. It does not do to blame the staff for mistakes, it is up to the supervision to take the management actions necessary to minimize the incidence of these errors. In doing this it must be recognized that the staff has only limited control over quality. It is the supervisor's responsibility to bring the staff to function at the level of the process capability. It is also his responsibility to develop methods to improve operations to better the production and quality of the department. To accomplish all this, the supervisor must take a very active role in quality control of his department.

Sampling

The QUIP system (see Figure 5.3) provides for a strong involvement on the part of the first line supervisor. After developing a list of errors which are to be checked, the supervisor checks the work of his staff to be sure that they are conforming to this list. The list has been found to be useful as a training tool as well as a guide for checking work.

Why is it that there is always

time to do it again but never time

to do it right the first time?

Figure 5.2

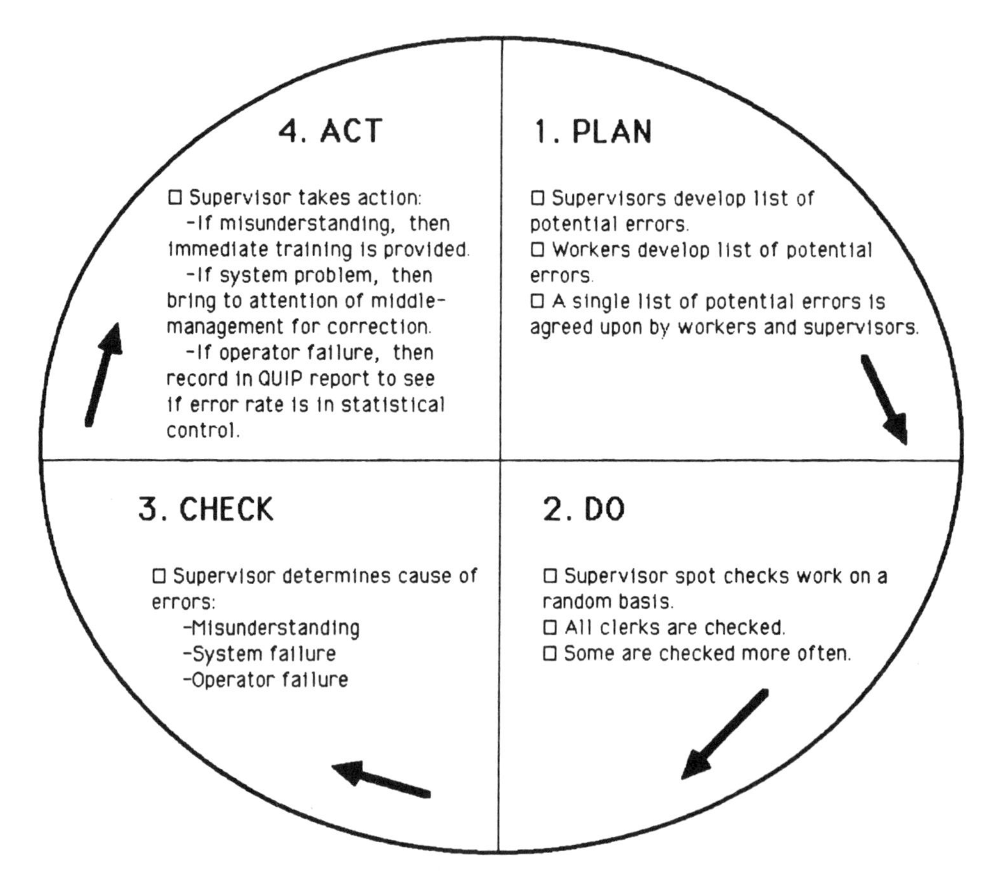

Figure 5.3 QUIP—A modern tool for supervisors.

The checks to be performed by the supervisor are not as rigorously scheduled as the checks made by clerical verifiers. In fact, the number of checks performed by the supervisor are not as important as that they occur frequently enough to make the workforce aware that quality is a necessary ingredient in production. It is recommended that the supervisor spend a portion of his time each day in making these checks. Twenty percent or about one and one-half hours a day are recommended minimum amounts [1].

The checks should not be all made at one time. They should be spread throughout the day in such a way that all employees are covered equally. They should be made at different times of the day, changing the sequence of which clerk is seen first, next, and so on. Employees should not be in a position to predict when they will be checked next. It is also good practice to return to some employees after having just checked them. This is done to prevent the employee from feeling that once they have been checked they can relax with regard to quality until the next day.

The supervisor must make it a point to check through the day even though this means that he will have to do this at pressure periods. It is not necessary at every such period when his attention is required in several places at the same time, but he must do it often enough to assure that the quality does not deteriorate at such critical times.

In examining the work of the clerks during his checks, the supervisor may come across error conditions. These obviously require immediate attention. Their treatment is described below.

How to Treat an Error

Since supervisors are generally very skilled in the tasks of the department they are often impatient with errors when they come across them. It is important, however, to reserve judgment and determine the cause of the error. This does not merely mean who committed the mistake but why it was made in the first place. Only in this manner will it be possible to effect corrective action. And only by determining the underlying causes is it possible to develop a long term correction program.

Causes and Actions

There are three primary causes for error. By identifying each of these causes, the appropriate corrective action can be taken. The three causes are (1) misunderstanding, (2) systems failure and (3) operator failure [2].

Misunderstanding—When a supervisor uncovers an error and takes this back to the clerk he is often asked, "Isn't this what I was supposed to do?" This type of statement, or similar expression, indicates a lack of understanding or training in handling the particular task. The supervisor should take the opportunity to explain the correct procedure then and there. This form of training is extremely effective.

A distinction should be made between this type of review and a review of a returned item. The item examined by he supervisor was just completed. The transaction was still fresh in the operators mind. The returned item, on the other hand, was completed some time past, perhaps days, weeks or even months ago. With the many transactions undertaken by the clerk, it is difficult to recall the circumstances of that particular item. As a result, an investigation into the cause for failure this long after the fact has little chance of uncovering a causal factor.

System failure—A number of errors which are uncovered are not under the control of the clerk at all. These are the system errors. They are often intermittent and sometimes appear innocuous. These are failures of the equipment, forms or procedures used in handling the transactions. When the supervisor discovers such errors he can either correct them at once or make the requests necessary to effect their correction.

Operator failure—The third possibility to be found on investigating an error condition is the operator failure. A simple goof. The treatment of this type of error presents more complexities than the misunderstanding or system failure. The former can elicit a reaction on the spot in the form of training or systems correction. The operator failure cannot be judged on a single observation. It makes a substantial difference if the observed failure was the first for the operator or the nth incident of the same thing. Such a failure can only be judged in relation to the process capability. Short of an immediate correction and ascertaining that no problem of training or system exists, the only sensible thing for the supervisor to do is to record the sample for the comparison with other records to the process capability.

Recording and Results

Whenever the supervisor examines the work of one of the staff, he records the result of the examination on a form similar to the one shown in Figure 5.4. This form remains with the supervisor for his use.

MANAGER'S SAMPLE

EMPLOYEE NAME					MANAGER'S NAME
A. MELENDEZ					T. RICE
DATE	TIME	ACTIVITY	WORK O.K.	WORK N.G.	REMARKS
6/12	10:00	TYPIST	10	0	
6/13	2:15	"	9	1	City of ORIGIN Not TYPED
6/14	12:55	"	10	0	
6/15	10:30	"	10	0	
6/16	2:15	"	10	0	

3242/00 (4-71)

Figure 5.4 Supervisor's record of spot checks.

With time, enough documentation exists to help pinpoint the cause of poor workmanship.

A separate form is prepared for each member of the staff. The form is updated every time the supervisor examines the clerk's work. In the example, an actual situation is represented [3]. The form has space for the clerk's name, the manager's name and data column. Date and time are recorded. The column marked "Activity" allows a record to be kept of the operation on which the clerk was judged. In small departments, clerks perform many different tasks of differing complexity. These tasks must be controlled separately since the process capability can only be developed on homogeneous operating conditions.

The two columns under the heading "Work" indicate the result of the sample. One column lists the number of acceptable items observed, the next column lists the number of errors observed. The last column shows the reason for failure.

This form is particularly valuable when a clerk is discovered to be out-of-control (i.e., so far removed from the process capability that the odds of that happening due to chance alone are less than one in three hundred). In such a circumstance, the data on the chart can be analyzed to uncover weaknesses in the clerk to be corrected.

The data for each clerk is consolidated on a weekly as well as on a cumulative basis.This information becomes the basis for the development of the process capability. The technical aspects of performing the mathematics of obtaining a process capability is reported in the literature of quality control [4–6].

Interpreting the Results

A weekly summary of the data collected by the supervisor is analyzed. A sample of such an analysis is shown in Figure 5.5 [7].

The salient features of this report are the process capability and the cumulative error rate or process average. There is a spread of nearly 0.7 percentage point between the two measures. This indicates a substantial room for improvement. The details of the report show that of the five clerks, clerk D represents the major contributor to the spread. Correction of this clerk would improve the process average.

Before undertaking to talk to the clerk, it is advisable to check the numbers used in the computation. An error in transcription could

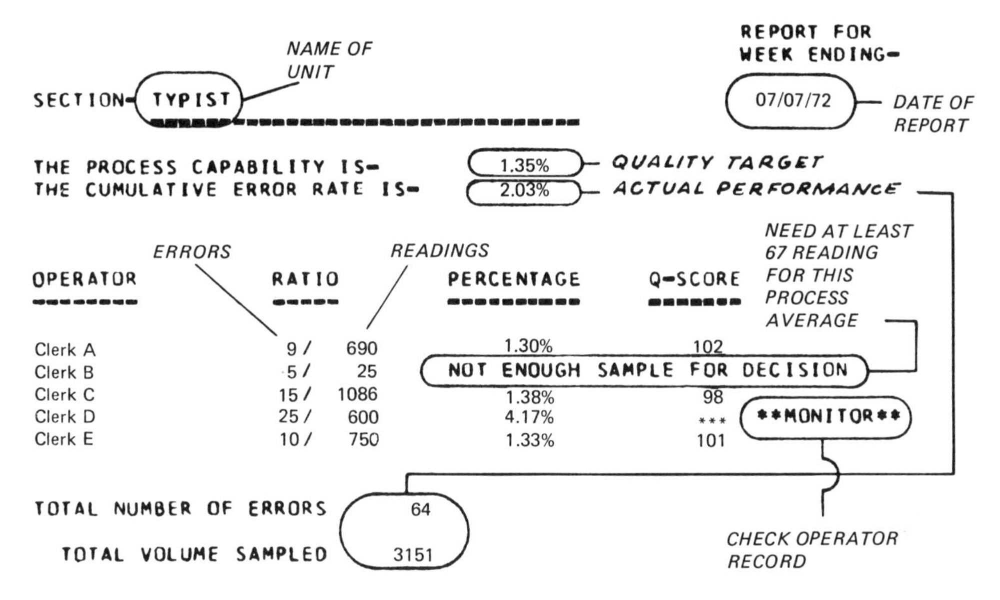

Figure 5.5 QUIP Analysis.

distort the result. If the arithmetic was correct, the next point to consider is whether this clerk performed the same work as the others. If this is not the case, then this clerk's data should be removed for separate treatment.

If the numbers are correct and the clerk does the same type of work as the others then it becomes important to determine the cause for this clerk's excessive number of errors. The manager's sample record (as illustrated in Figure 5.4) is consulted for clues to the clerk's problem. An analysis of when the errors occurred (day of week or hour) can often reveal a trend. A breakdown of error type can disclose specific problems. It should also be considered whether the clerk has any physical deficiencies which would prevent achieving top quality. For instance, if the clerk needs glasses or a new prescription, obtaining these will alleviate many problems. External causes related to where the clerk is working can also contribute to errors. The discovery and solution of these problems is the responsibility of the supervisor.

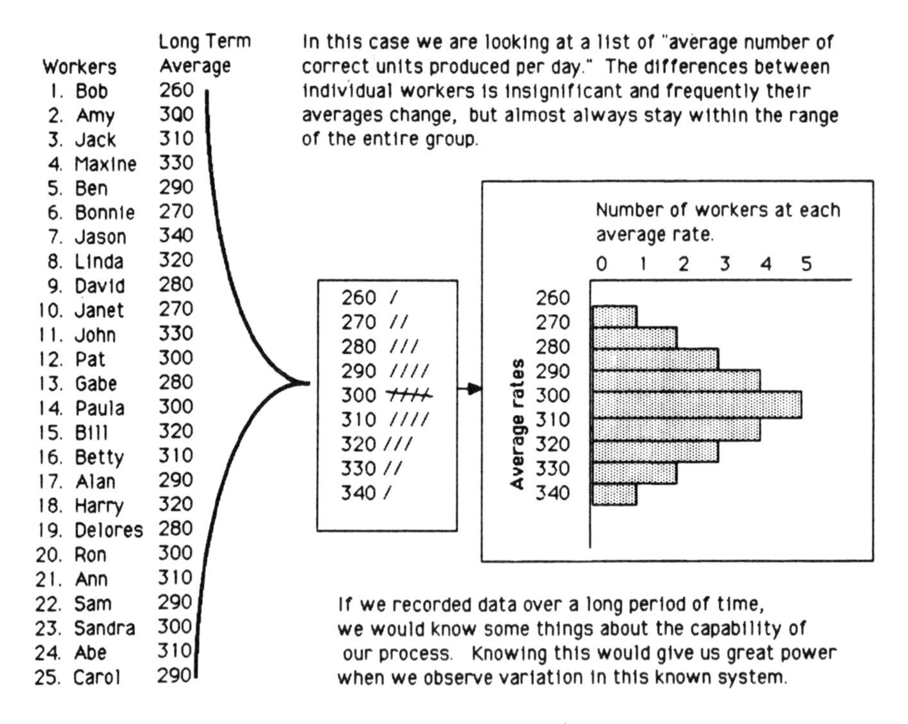

Figure 5.6a Process capability: setting up the measurement.

However, with the data available to him this should not be an insuperable task.

The Process Capability

The principle of process capability is most significant in quality production. In essence it states that there is an error rate that is inherent in the system. The best efforts of the employees cannot do more than to achieve the process capability. Only management can improve the system to improve the process capability. (See Figure 5.6.)

The system is taken in its broadest interpretation when used in

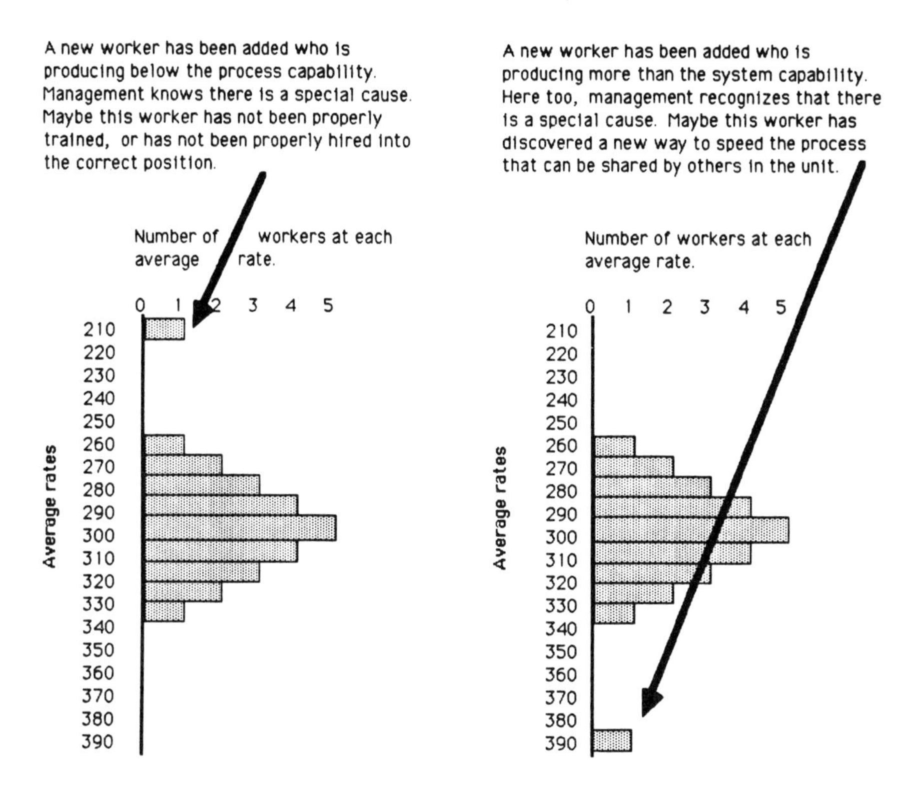

Figure 5.6b Process capability: identifying special causes.

context of the process capability. It refers to all the environmental and causal factors that make up the operation being considered. Thus the training and caliber of clerks make up part of the system. So do the lighting, heat, workflow and many other factors. (See Figure 5.7.)

Many problems of quality are due to a faulty system. As Deming states

> Another purpose [of this paper] is to point out to management that most of the trouble with faulty product, recalls, high cost of production and service, is chargeable to the system and hence to management. Effort to improve the performance of workers will be a disappointment until the handicap of the system is reduced [8].

HUNDREDS OF VARIABLES ARE CONSTANTLY ACTING ON EVERY PROCESS. THE CHALLENGE IS TO BEGIN BY REMOVING ALL SPECIAL CAUSES.

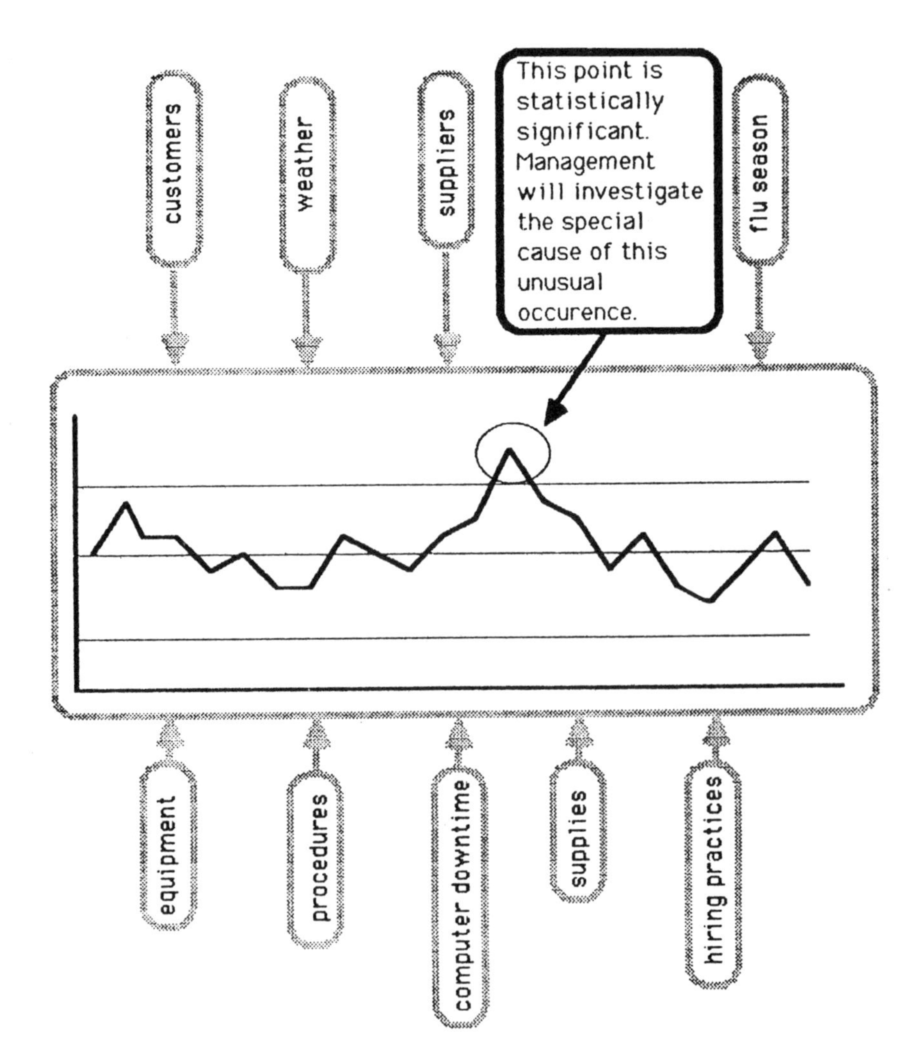

Figure 5.7 Many factors influence the system.

At the 1974 American Bankers Association Operation Conference a similar statement was made:

> It is very important to distinguish between [common and special causes]. The Quality Improvement Program works by determining the common causes (process capability) then isolating those operators who do not operate within this capability. We find from our experience that approximately 75% of the error rate is due to common type of causes which can only be removed by some managerial or systems change. The other 25% of the error rate is due to operator type (special) causes [9].

Quality control can identify and segregate the types of errors by cause. With that management can take effective action to improve quality.

References

1. Latzko, William J., "QUIP — The Quality Improvement Program," in *Twenty-Ninth Annual Technical Conference Proceedings* (San Diego: American Society for Quality Control, 1975) p. 249.
2. Latzko, William J., "QUIP — The Quality Improvement Program," in *Twenty-Ninth Annual Technical Conference Proceedings* (San Diego: American Society for Quality Control, 1975) p. 248.
3. Latzko, William J., "QUIP — The Quality Improvement Program," in *Twenty-Ninth Annual Technical Conference Proceedings* (San Diego: American Society for Quality Control, 1975) p. 250 and 253.
4. William J. Latzko, "Stabilized t-Charts, Theory and Practice," in *Twenty-Third Annual Technical Conference Proceedings* (Los Angeles: American Society for Quality Control, 1969).
5. William J. Latzko, "Clerical Process Capability," in *Twenty-Fifth Annual Conference Proceedings* (New Brunswick: Metropolitan Section, American Society for Quality Control, 1973).

6. William J. Latzko, "Basic Tools for Clerical Quality Control," in *Twenty-Seventh Annual Conference Proceedings* (Middlesex, Edison, N.J.: Metropolitan Section, American Society for Quality Control, 1975).
7. Latzko, William J., "QUIP — The Quality Improvement Program," in *Twenty-Ninth Annual Technical Conference Proceedings* (San Diego: American Society for Quality Control, 1975) p. 256.
8. Deming, W. Edwards, "On Some statistical Aids Towards Economic Production," *Interfaces*, Vol 5, No 4 (August 1975) p. 1.
9. Latzko, William J. "Quality Control in Banking," in *The 1974 National Operations and Automation Conference Proceedings* (Washington: American Bankers Association), p. 42.

6
The Economics of Quality

Every bank, whether it has formal quality control or not, is spending substantial sums to maintain quality. "It is estimated that bank spend approximately eight percent to ten percent of their sales or operating income on quality costs, and that 25 percent to 40 percent of the bank's labor costs are associated with quality costs" [1]. With the large sums involved, it is worthwhile to examine the concepts of quality costs and how they can be applied to optimize a bank's profits.

The Components of the Costs of Poor Quality

It is generally agreed that there are four elements of costs related to quality that lend themselves to measurement:

1. Appraisal
2. Internal failure
3. External failure
4. Prevention

These cost elements make up the major segments identified by the American Society for Quality Control [2]. These elements are further subdivided to identify the components so that accountants can incorporate them into their costs systems. These components generally relate to manufacturing concepts but were recently transcribed to be used by service industries such as banking [3].

In addition, there are at least two other economic factors related to quality that cannot be readily measured. These components are:

1. The customer's satisfaction
2. The employee's satisfaction

These elements are probably far more important than the measurable ones. (See Figure 6.1.) In fact the concentration on measurable costs is probably misleading. The key to successful quality management is not the concentration on costs but on profits [4].

The measurable costs related to poor quality are of some importance in that they demonstrate the visible losses of poor quality. As long as they are taken in that context they can be helpful. In this type of use, the measurable costs give rise to the use of process control to reduce failures in the first place. The reader is cautioned against trying to use the measurable costs to minimize the total cost. Such action can suboptimize the system, reduce profits and lead to corporate loss even though there are departmental savings.

In the proper context it is worthwhile to examine the detailed description of the measurable costs of poor quality.

Appraisal

The cost associated with determining that the work conforms to the established standards is called the appraisal cost. This is the cost of checking, verifying, signing or any similar activity. It is a nonproductive labor cost that exists merely to be certain that the original work was done correctly.

In clerical operations, the function of appraisal is important and often an unavoidable cost. There are, however, techniques available for making the appraisal function more cost effective. Methods of stratification and batch control have been reported [5].

MEASURABLE COSTS

CATEGORY	DEFINITION	EXAMPLES
1. Appraisal Costs	☐ Determine that work conforms to a standard.	• Loan approval • Checking manifolds • Verification of funds transfer advice
2. Internal Failure Costs	☐ Fixing that which get's caught before going to customers.	• Reject repair • Re-writing checks • Re-working loan agreements
3. External Failure Costs	☐ Fixing that which gets caught by customers.	• Penalty costs for money transfer errors • Adjustments for mis-posting • Bad loans
4. Prevention Costs	☐ Investment in minimizing the above three costs.	• Training • Development of quality control systems • Management teams working to improve quality

UNMEASURABLE COSTS

CATEGORY	DEFINITION	EXAMPLES
1. Customer Costs	☐ Loss of business	• Customers reduce volume of business • Customers switch banks • Customers complain to friends
2. Worker Costs	☐ Loss of productivity	• Malaise

Figure 6.1 Examples of cost categories for quality control studies.

An interesting aspect of the appraisal cost is that it often extends fairly high in the corporate ranks. Loan reviews are, to a large extent, a form of appraisal. They are not normally thought of as a cost of quality, yet, strictly speaking, they fall into this category. Stratification is often applied in this form of appraisal: the number and level of reviews depends upon the size of the loan request.

Internal Failure

When the appraisal function detects an error, the work is rejected.In most cases, it has to be reworked. The costs associated with performing the work badly in the first place, the cost of forms, the cost of computer time, the related costs of operation and the cost of redoing the work (instead of a new item) are all part of the internal failure cost.

Internal failure is, perhaps, the least controlled element of cost in a bank. The clerk who mistypes a check, notices the error, takes it out of her machine, tears up the check and types a new check is an example of internal failure costs: all the components for this cost are present. Many cost studies allow for internal failure. They build the cost of bad quality into the system and perpetuate it. What is even more disturbing is that most cost studies use the process average which nine times out of ten greatly exceeds the process capability. This allowance for bad work denies managers any cost incentive for quality. Only when the system deteriorates beyond the already bad level of quality does a problem become obvious.

External Failure

If the mistake is not caught in the appraisal function, it leaves the department to go to the next unit or even the customer. When it is ultimately returned, the correction of this type of error costs as much as the internal failure and often more. Because of a time lapse between doing the work and its return, research is often required. In addition, adjustments and penalties must frequently be paid. These are out of pocket expenses which are highly visible. Also, there is an impact on the customer which cannot be readily quantified.

Because external failures are recorded separately and involve direct expenses to the bank, they are the recipient of much management attention; not always in the best way.

Frequently, management will overreact to external failure. Rather than systematically and scientifically attacking the cause of the failure, management will add further appraisers increasing both appraisal and internal failure costs without coming to grip with the real problem. Often the appraisal cost and internal failure cost are raised beyond any possible savings in external failure costs [6].

External failures are controlled by the nature of the costs. If similar controls were also applied to appraisal and internal failure costs, a balanced approach to quality cost and quality control could result.

Prevention

The costs of quality planning, quality training, quality audits, design and development of quality measurements and test equipment, as well as the normal costs of a quality control department, such as supplies, are the cost of prevention. This represents the investment that management makes in minimizing the total costs of quality.

Analyzing the Costs of Quality

The total cost of quality is the sum of the four elements. Since these elements are related, changing one impacts the others. As a result, the total cost curve tends to be shaped like a parabola. Theoretically, it should be possible to develop a set of conditions relating the quality cost elements in such a way as to minimize the total cost of quality.

Most attempts at formulating a general model bog down in the complexity of the probability mechanisms associated with the components of the cost elements.

While the general model eludes the investigator, it is possible to develop an analysis to solve specific, practical problems. It is possible, for instance, to determine the distribution of the costs elements in relation to the total cost. Table 6.1 shows a reported distribution. At the same conference where this was presented, several other papers were read showing remarkable similarity in their distribution [6].

The data from Table 6.1 shows the relative magnitude of the uncontrolled cost functions in relation to the total cost. From this it would appear that the problem of quality in the banking industry

TABLE 6.1 Distribution of Quality Costs

1.	Appraisal Cost	28%
2.	Internal Failure Cost	41%
3.	External Failure Cost	29%
4.	Prevention Cost	2%
	Total Quality Costs	100%

Source: William J. Latzko, "Reducing Clerical Quality Costs," in *Twenty-Eighth Annual Technical Conference Proceedings* (Boston: American Society for Quality Control, 1974), p. 186.

should be approached from the cost reduction point of view. The costs of quality should be isolated so that managers can be made responsible for them. In this way, managers will have the profit incentive to improve quality, a function which is lacking in the industry today.

Inspection Costs — An Example

The example presented here brings together some of the elements of the QUIP discussed in Chapter 5 and the concepts of quality costs discussed above. The case is based on actual experience although the data is somewhat coded.

Statement of the Problem

The department involved handles numerous transactions daily which range in dollar value from less than ten dollars to over one hundred millions dollars. The average transaction is about $30,000.

The processing is essentially a manual function involving a clerk who prepares the item for a typist to type. The transaction is then checked for accuracy and finally a "signer" initials the item after checking it once more.

The care exercised in performing the work is great. The error rate escaping to customers is less than 0.7% of the work volume. Although there is less than one error in 1,500 transactions, the cost of these errors is very high, running in the hundreds of thousands of dollars annually.

The QUIP system was installed and indicated that the operators were functioning at the level of the process capability. To improve the situation a change of the system was required. A small number of errors were costing a great deal. To find these errors was difficult since there were so few in relation to the work.

Data Sources

The QUIP system furnished information on the operating rates of each of the clerical units including the checkers. The external error rate was obtained from an analysis of the adjustments which were necessary to correct the items returned as erroneous by the customers. Even though the data may not have been as accurate as one might desire, from a practical point it was very adequate. An analysis of the distribution by dollar amount of both the incoming work and the errors was also available. This analysis quickly showed that the main cost of external failure was in the very large dollar amounts.

Analysis

Since the internal failure rate was already in a state of control and, short of a long term computerization of the activity, no real system changes suggested themselves, it was necessary to look elsewhere for a solution.

The obvious place was in the checking function. However, since the documents were already checked twice, would additional levels of checkers pay for themselves? To determine this it was necessary to analyze the efficiency of the appraisal function, develop the probability of failing to find a defective item after the nth level of checker and use this probability to evaluate the total cost of quality for each of n levels of checkers. The choice of level of checker would be dictated by the optimum cost of quality.

To develop the relationship needed to arrive at the proper choice it is necessary to consider, in sequence, the efficiency of more than one checker, the odds that a defect will occur before checking and, from this, the probability of not finding a specific defect. The formulas are derived in Appendix 2.

Applying the formulas to the actual data showed that the checkers found approximately 75% of the defects. The work as presented to the first checker was about 1.25% defective. With sufficient appraisers to check every item, the escape rate, defective items not detected, would reduce to less than one in 5,500. This is, of course, theoretical limit of checking efficiency which relies on the fact that all the assumptions of Appendix 2 will hold. Since the first assumption of independence between checkers is weak, it was felt that the theoretical number was too optimistic. Applying the experience of the existing operation it was felt that an error rate of about one in 2,800 was more realistic.

A further analysis disclosed that, while the bulk of the adjustments related to items with a dollar value over $50,000, this class of transaction accounted for only 40% of the total volume. This segmentation of the transactions allowed the use of fewer third level checkers. An analysis of the total cost of quality disclosed that the addition of the third level of checkers to examine work over $50,000 would save an incremental $80,000 per year. (See Figure 6.2.)

Implementation—Since only 40% more checkers were needed, they were put on an alternate shift. This allowed the checkers to have easy access to the tools needed to do their work and alleviated pressure situations for them. Working in quieter surroundings, they are able to perform better than the regular shift checkers. The increased efficiency that resulted from these moves allowed them to detect more errors than forecast. Unfortunately, the restraints of the operation are such that it was not possible to recreate the same good checking conditions for the regular shift checkers.

Three years later—The program was so successful that more than the expected savings were realized. As part of the quality control reporting system, a record was kept, and is still maintained, on the error rate as reported by each of the three levels of checker. Since the operation was in a stable condition when the problem first arose, quality of the work showed only random fluctuations. The report of external failure rate also reflected the addition of the third level of checker. It dropped off as the program was implemented and remained stable.

The problem of the internal failure rate was tackled and a computerized method of handling the transactions is now partially implemented. This system will allow savings to be realized by the reduction of the appraisal units as the automation takes place. Some early results indicates that the forecast savings will be achieved.

STARTING SITUATION

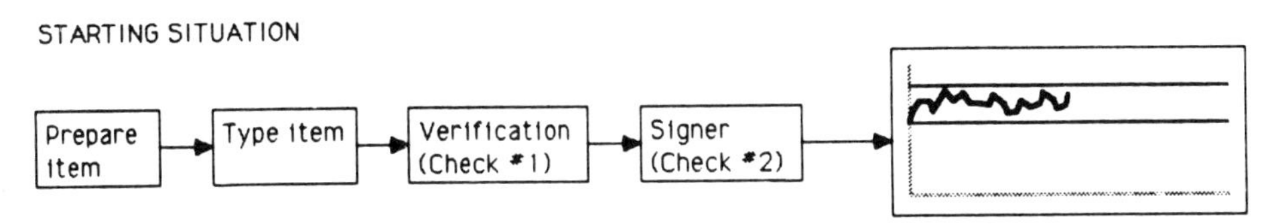

In this situation, the error rate of 1 per 1,500 items, was well established over time. However, each error involved hundreds of thousands even millions of dollars.

IMPROVED SITUATION

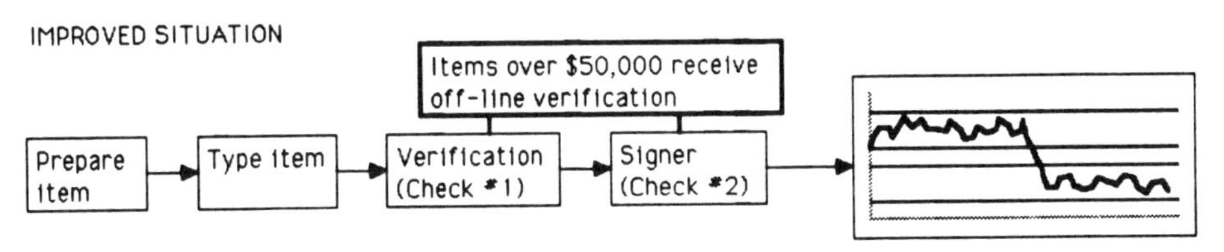

Study showed that the bulk of errors occured in items over $50,000. So these were given special attention off line. The error rate went to 1 per 5,000 items, saving the bank $80,000. per year.

Figure 6.2 An example of an improvement project.

Observations

A knowledge of quality cost can be very helpful in bringing about improved quality control techniques and evaluate the effects of new systems. When the risks can be quantified, it is possible to weigh the benefits of several alternatives to choose the optimum. However, one should not be blinded by a cost analysis. Such analyses assume a static process which is usually not quite true. Built into the "Quality Cost" is an implicit assumption that appraisal is the main protection to quality and is 100% accurate. Nothing could be further from the truth.

The model does consider the prevention costs but these fall short of what is really needed to achieve quality, process controls. The motto should be: "inspect the process, not the product." (See Figure 6.3.) Process control can achieve increased levels of good quality far more economically than appraisal. Process control is dynamic and modern. Appraisal is static and out-of-date.

Trying to balance costs using appraisal is futile and can easily be counter productive. In effect it assumes a level of quality exists which is optimum. The only such level in a competitive environment is

Inspect the process,
not the product.

Figure 6.3

perfect quality. Anything less can cause consumers to switch from lower quality to higher quality despite price. Professor Taguchi has developed a sound theory along these lines which was validated in a practical way by the Ford Motor Company's Batavia plant experiment.

Similar findings come from consumer surveys. In a survey conducted by The Gallup Organization, Inc. for the American Society for Quality Control they report:

> On average consumers report they would pay about a third more for a better quality car ($13,581 rather than $10,000). Consumers would be willing to pay about 50% more for a better quality dishwasher ($464 rather than $300), and proportionally more for a television or sofa they thought was of better quality ($497 rather than $300 for a TV, and $868 rather than $500 for a sofa). Finally, consumers claim they would, on the average, pay twice as much for a better quality pair of shoes ($47) than for an average quality pair ($20). [7]

The point is that cost savings are far less important than quality to consumers in a competitive environment. That this also holds for service industry such as banks was found by a Citibank study that showed the same relationship for bank products as the Gallup survey showed for durable items [8]. It is profit not costs that are important. And quality cannot be gained through cost savings, it is gained through the application of good management methods as outlined in Chapter 9.

References

1. Latzko, William J. "Quality Control in Banking," in *The 1974 National Operations and Automation Conference Proceedings* (Washington: American Bankers Association), p. 38.
2. American Society for Quality Control, *Quality Costs—What & How*, 2nd edition (Milwaukee, Wisconsin, 1971).
3. William J. Latzko, "Quality Cost for Service Industry," an unpublished report presented by Task Force II of the Quality Control Cost Committee of the American Society for Quality Control at the Annual Technical Conference, Philadelphia, May 12, 1977.
4. Kume, Hitoshi. "Business Management and Quality Cost: The Japanese View," *Quality Progress*, Volume 18, Number 5, May 1985, pp. 13-18.
5. William J. Latzko, "Reducing Clerical Quality Costs," in *Twenty-Eighth Annual Technical Conference Proceedings* (Boston: American Society for Quality Control, 1974), p. 188.
6. Although several papers were presented relating to Quality Cost, only one other published distribution data: A. F. Grimm, "Quality Costs: Where Are They in the Accounting Process?" *Twenty-Eighth Annual Technical Conference Proceedings* (Boston: American Society for Quality Control, 1974), p. 198.
7. The Gallup Organization, Inc. "Consumer Perceptions Concerning the Quality of American Products and Services," *American Society for Quality Control*, September 1985, p.12.
8. "Poll Finds Consumers Unhappy on Quality " *American Banker*, September 22, 1976, p. 7.

7

Organizing for Quality Control

In discussing "Road-Blocks to Quality in America," Dr. W. Edwards Deming makes the following important statement:

> An obstacle that ensures disappointment is the supposition all too prevalent that quality control is something that you install, like a new Dean, or a new carpet, or new furniture. Install it and you have it. This supposition is unfortunately force-fed by the common language of quality control engineers, some of whom offer to install a quality control system. Actually, quality control, to be successful in any company, must be a learning-process, year by year, from the top downward and from the bottom up, with accumulation of knowledge and experience, under competent tutelage [1].

Unfortunately, the concept of what quality control is or what it should accomplish is seldom considered in organizing to achieve quality. "The basic function of all quality control activity is to provide information which will be of assistance to those who are responsible for the various operations which combine to produce the product" [2].

Quality control of some form or other is carried on in all banks. Those banks which wish to optimize this function and, thereby, reduce their cost of quality find it useful to create a function or even a department. As Dr. Juran stated, "Creation of a new department [or function] should be accompanied with preparation of a statement of the objectives of the department, and a plan for carrying out those objectives" [3]. Dr. Juran further recommends that this be done in cooperation with other affected departments to assure sufficient participation by all in the quality process so that the measurement and staff functions of quality control have a reasonable chance to succeed.

Most authors agree that a quality control function cannot succeed unless it has the full, active support of senior management. "The responsibility for a successful Quality Program, however, does not lie entirely with the Quality Control Department. By virtue of its position in a Company, management has the basic Quality Responsibility [4]." Mr. Lieberman goes on to enumerate the responsibilities of management to assure a successful program:

1. Delegation of sufficient authority for the department to be effective
2. Providing sufficient managerial information for the department to contribute to policy
3. Provide the resources of personnel and equipment
4. " . . . accept that department as a participant in all discussion and decisions related to quality" [5]

Once a management is committed to expanding the resources necessary to achieve the economic control of quality, they are faced with some practical questions such as the form that the function should take, the responsibilities and authorities to be given and how to best implement the decision.

Quality Control Organizations in Banks

In general, banks which have a formal quality control program are operating either on a centralized, decentralized or hybrid form of organization structure [6]. The choice depends on the size of the bank and the degree to which the program has been formalized. Although it is not true in every case, the less formal the quality control structure, the more likely it is to be decentralized.

Centralized Control

A very formal, structured quality control system might use a centralized organization. This is the type of organization often found in manufacturing. This type of organization has a staff quality control department whose inspectors do all the measurement and whose analysts work under the direct control of the quality control management. The advantages of this type of organization are the ease with which control can be exercised, the reliance which can be placed on the data which is collected by the organization and the standardization of the management information system that quality control generates.

Among the disadvantages of such a system are the lack of line cooperation ("them" against "us"), the expense of controlling a large staff of inspectors and the feeling of intrusion that members of another department and staff area give in an operation. This is particularly the case if the management of the quality control function is inept, giving the sense that their sole duty is to act as a policeman to catch the culprit who made a mistake rather than acting as a professional who is trying to assist the department in achieving optimum quality.

Decentralized Control

Decentralized control is often encountered in small organizations and those which do not have a formal quality control organization. The advantages of this type of organization are related to the disadvantages of the centralized system: this form of control tends to be less costly [7], generates a sense of involvement on the part of the operation and generates more cooperation than a staff function might obtain.

The disadvantage of decentralized or line control is related to the advantages of the centralized system: Any data generated is often viewed with suspicion of being less than accurate, data collection methods are often not uniform and reports are generally not standardized. Another problem of great severity with the line control is that in periods of stress, the control is eliminated when most needed. In one large bank, decentralization of the quality control function resulted in the analysts being used as trouble shooters and the gradual elimination of what had been an effective, cost saving system. "Staff departments who persist in devoting their energies to day-to-day problems thereby not only miss the greater opportunities available through solving chronic problems; they build up resentment in the mind of the line people because of the day-today interference" [3].

Hybrid Control Systems

Successful quality control organizations are a combination of the centralized and line form of control. They avoid most of the disadvantages and gain many benefits. In general, the organization of these banks has the inspection or appraisal function under the supervision of the line personnel with the necessary report and audit controls to assure compliance to efficient quality control procedures. The staff areas in this type of an organization act as the policy makers with regard to quality and as the professional staff which established the procedures needed to achieve optimum quality in the most economical way.

Responsibility and Authority

"It must first be recognized that the control of product quality is not the exclusive responsibility of the quality control organization . . . the quality control function [is] to measure the quality performance of the participating groups and to provide protection against unacceptable performance by identifying deficiencies and enforcing the necessary corrective action" [8]. Quality control is not responsible for the quality of the product or the design, that is the responsibility of the operating management. Quality control is responsible for determining how well the product conforms to the specification and if it is not acceptable to insist that the operating management take the necessary action to correct the deficiency.

Quality control is also responsible to inform senior management of the level of quality attained by the operation, the costs involved and develop standards to enable management to judge the adequacy of the quality level performed by the operations areas. The management reports should be developed in such a way that management can determine quickly whether the overall level of quality is in control or not. Reports of this type are also sales tools in that they can be used to demonstrate to prospective customers the high levels of quality achieved by the bank.

In many ways, the responsibility of quality control is similar to that of auditing. Like the auditor, the quality control manager must assess the operation from a quality viewpoint and report any deficiencies which are uncovered. It is up to the operating management to remove deficiencies. Like the auditing function, quality control's responsibility covers all aspects of the bank. Consequently, it would be a mistake to limit the activities of quality control to only one area of the bank such as checks processing.

Implementing a Quality Control Program

Once a management decision has been made to establish a formal quality control program and the scope and responsibility of the program are developed, it is necessary to staff the function. The essential need is to have someone designated to be responsible for the program. This should be an officer of the bank. In a small bank this may be an officer who has this task in addition to other duties. In a larger organization it may be specialist function only in the area of quality control.

Although a bank officer should be in charge of the quality control function, there are three approaches that can be used to get the necessary technical expertise:

Hire an experienced quality control expert and teach him banking operations.

Take a banker with a predisposition as a cost oriented problem solver, and teach him quality control. To this end several colleges including the University of Connecticut and Purdue University, as well as the Extended Training Institute of the American Society for Quality Control, offer courses. There are also a variety of standard textbooks on quality control. These, however, have an industrial orientation and a degree of experience is necessary to convert the industrial art into a commercial banking reference.

> The final method, and perhaps the best for a bank exploring the opportunities of quality control, is to seek the assistance of a consultant. The American Society for Quality Control will supply a list of consultants in your area. [9]

Most of the successful applications of a formal quality control system in banks have been implemented by quality control experts who came from some background other than banking.

It has been found useful to start a quality control program on an adequate but small scale and use the savings generated from the program to further expand it.

> *Start small.* Irving Trust has some 5,000 employees and deposits of nearly $7.4 billion. Yet when it established its Quality Control Center, only two clerks and the author were involved. Within two years, application were increased tenfold, but only three analysts were added. Since then, others have been taken on. The point is, the increases have been made slowly, as the need arose. You can do the same.
>
> *Get top management support.* This is an absolute must. Otherwise, a program may wither for lack of interest. In obtaining such support, management must be persuaded to plough back some of the savings the program engenders into new programs. These savings will provide a simple criterion by which to measure the worth of the effort. Furthermore, they will eventually show up on the bank's bottom line. [10]

The success of a quality control program is directly related to the management support engendered and the skill of the quality control department.

References

1. W. Edwards Deming, "On Some Statistical Aids Towards Economic Production," *Interfaces*, Volume 5, Number 4 (August, 1975), p. 2.

2. William L. Lieberman, "Organization and Administration of a Quality Control Program," *Industrial Quality Control*, January, 1962, p. 27.
3. Joseph M. Juran, "Organization for Quality" in J. M. Juran (Editor-in-Chief), *Quality Control Handbook*, 2nd Edition (New York: McGraw-Hill Book Company, 1962), p. 6-29.
4. William L. Lieberman, "Organization and Administration of a Quality Control Program," *Industrial Quality Control*, January, 1962, p. 29.
5. William L. Lieberman, "Organization and Administration of a Quality Control Program," *Industrial Quality Control*, January, 1962, p. 29-30.
6. Much of the concepts of organization are taken from a lecture by Michael P. Quinn, AVP and Director of Quality Control at Manufacturers Hanover Trust Company, presented at the American Institute of Banking in New York, January 2, 1978.
7. Line control may have a lower cost of appraisal but that is by no means certain. It would require closer cost controls than exist in most banks to determine if the fragmentation of quality control really results in a lowered cost quality. Duplication of effort and less sophistication could actually increase costs.
8. John T. Hagan, *A Management Role for Quality Control* (New York: American Management Association, In., 1968), p. 24-25.
9. William J. Latzko, "A Quality Control System for Banks," *The Magazine of Bank Administration*, November, 1972, p. 23.
10. William J. Latzko, "Quality Control for Banks," *The Bankers Magazine*, Volume 160, Number 4 (Autumn, 1977), p. 67.

8

Starting a Quality Program

When a bank first becomes interested in the quality control process there are two important questions facing them:

1. How do we organize to do this job?
2. Where do we begin?

The first question was addressed in the chapter on organization. The second question will be covered in this chapter.

Although it is possible to start in more than one area, that is not common. Usually, there are not enough resources available to do every thing. Besides, it is a good idea that the system be "sold" by showing its value in application. Top management finds that there is more bank-wide acceptance of the concept of managing quality if application in an important department demonstrates the value of the method. Normally, a significant area of the bank is chosen as the place to start.

There are a number of factors in selecting the area to begin. Usually, management has some department in mind for the implementation. If there is no agreement of where to start, it is advisable to list the bank's losses from various sources such as from loan write-offs, security transactions, funds transfer operations, and checks processing operations, among others.

By ranking this list in decreasing order of loss, the bank can see where the out-of-pocket costs are largest. This may or may not be a good perspective on the actual losses to the bank. After all, the inability to get loans due to poor management may be a much greater loss to the bank than any measurable out-of-pocket amount.

While it is not possible to specify starting in a particular department—each bank being different—experience provides several areas which are good starting places: commercial lending, securities, funds transfer and checks processing. These are by no means the only possible regions. Personnel, computer systems, mortgage lending, and retail banking are good potential starting points.

Once the domain is selected, the task is to determine what to do in that area. One method of determining what to do, the Quality Measuring System (QMS), is based on the work of Adam, Hershauer and Ruch. They are the authors of a National Science Foundation study entitled, "Measuring the Quality Dimension of Service Productivity" [1].

The authors are concerned with productivity measures. On the title page of their manuscript they show the economic formula for productivity, O/I = P: Output divided by Input equals Productivity. The formula is easy to understand. The problem has always been how to measure both he output and the input. Adam *et al.* evolved a structured method to generate a set of output and input measures. The QMS uses their technique but produces quality rather than productivity measures.

Although Adam *et al.* developed the system in banks, and although the Bank Administration Institute sponsored further research by the team, there is little evidence that the system of quality/productivity measures were successfully used by banks. The trouble appears to be that the productivity measure is basically dimensionless and only indirectly related to factors under the control of management. As a result, these two-dimension measures do not give clear indications to managers what to adjust. Nevertheless their work is of great significance since the steps just prior to creating the productivity measure are the same steps used in the QMS. The difference comes about in the type of measure formed (QMS is entirely quality related) and the step in developing the plan for measurement.

Before examining the QMS, it is worthwhile to consider the Quality Productivity relationship.

The Quality Productivity Relationship

When quality goes up, productivity goes up. Why? Because there is less rework. The time spent on rework (or correcting a problem) is lost time in two ways. First the actual production time is lost. Secondly, with fixed or limited resources, the rework time cuts into the productive time. Rework is non-productive labor. When a loan officer is sent back to get more information that he should have obtained in the first place, the extra time involved is wasted labor. (See Figure 8.1.)

Another consideration, often overlooked, is the impact on the workforce. Rework interferes with the orderly flow of processing and so reduces the productivity even further. When people who are posting loan applications come across an illegible or missing data item, they must contact the originator of the application. The originator is often not immediately available. The work is put aside and other work done while waiting.

When the originator calls back with the answer, the clerk must pickup from where he left off. This often requires the re-reading (and re-computing) of data to get it right. Not only is the opportunity for error increased but extra time is consumed to "get back on track."

By eliminating the causes of error, wasted time is eliminated. **When quality goes up, productivity goes up.**

The Theory of QMS

QMS is a systems approach to defining what to measure. It applies some principles of group dynamics that set the stage for later work. It has been useful in giving the people involved a better sense of their operations as well as an insight into a problem area that was not as clear before. Because the result of the process is a series of quality measures developed by those who are involved in the operation, the measures are accepted and solutions are frequently suggested by those who might otherwise resist or resent the implication that their department is less than perfect.

THE QUALITY/PRODUCTIVITY RELATIONSHIP

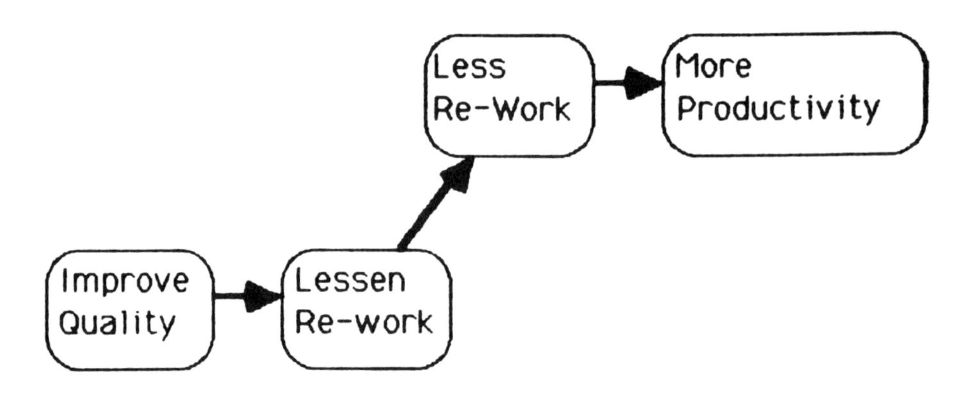

"It just makes sense..."

- Rework interferes with the orderly flow.
- Items needing re-work are often set-aside for special attention.
- Time, energy and money is lost because re-work requires:
  - ☐ re-reading,
  - ☐ re-computing,
  - ☐ searching for correct information, and
  - ☐ extra supplies

Figure 8.1 Productivity improvements follow quality improvements.

Systems Approach

The first step in the QMS is to clearly define the process about to be studied. That sounds simple and is simple if approached in a systematic manner. To define the operation clearly, a "system boundary" is established. (See Figure 8.2.)

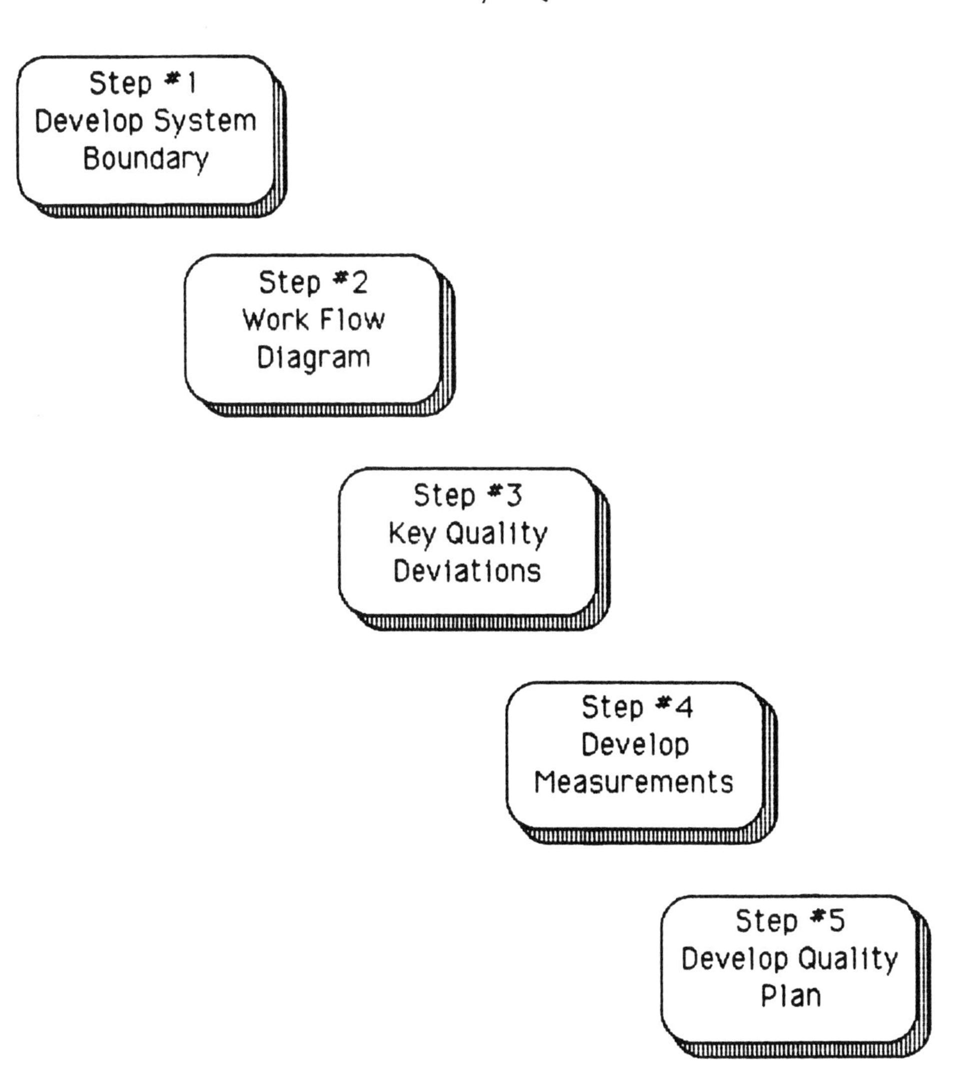

Figure 8.2 Steps in the quality improvement process.

System boundary—The concept of the system boundary is to define the limits of the operation, what goes into the operation, what comes out of the operation, what is needed to make it work and who interacts with the operation. (See Figure 8.3.)

In effect, one can think of all operations as receiving raw material

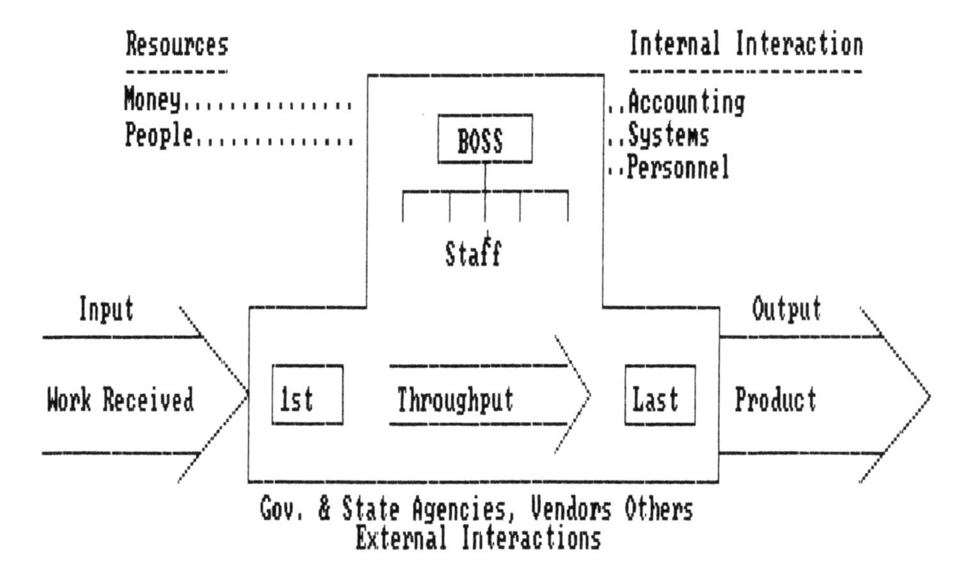

Figure 8.3 Example of system boundary.

and delivering a finished product (as far as the operation is concerned). With a little thought it is possible to define the raw material (input) of a clerical or professional process. For instance, the raw material for the recruiting operation of a personnel department can be thought of as the request received for a given number of employees with given skills. For a loan process it is the receipt of a loan application.

The product (output) can be developed in a similar manner. For the recruiting operation of a personnel department it can be thought of as the acquisition of a suitable person who will remain on the payroll a minimum length of time. In the loan process it can be a posted loan. That which is the output of the department is the product that results from the process.

The input and the output are on the outside of the system boundary (see Figure 8.3). There is no direct control by the department of what happens outside the system boundary. There may be influences but these are indirect.

Within the system boundary there are two forces: 1. the work to be done in converting the input to output and 2. the organization that accomplishes task 1. These forces are sketchily described within the system boundary.

Outside the boundary are other factors that influence the operation. One of these are the resources with which the department works. The resources generally consist of people, equipment, and supplies, among others.

Interactions are the other factors. There are two types influencing the workings of the operation: 1. internal interactions and 2. external interactions.

The internal interactions are the relationships that a department has within the organization. They are the common areas of personnel, accounting and a host of staff and operational areas. These areas influence the work by sending inputs or receiving outputs. Their function is either to support or be a more involved activity. Whatever their function, they are part of the organization although not a part of the system.

External interactions come from groups outside the corporation. For instance, government agencies control much of what goes on in a bank. In addition, vendors of equipment and supplies have a great deal of association with the department's activities. Other groups such as personnel recruiting agencies, temporary help agencies, collection agencies and consultants among others can have a direct influence on the work.

The basic purpose for clearly defining the system boundary is to create a map of the area of responsibility for the department. What is done inside the system boundary is the department's responsibility although other factors outside the boundaries may have an influence. The task of defining the quality measures is limited to that which relates to the input, throughput and output. Other portions of the task, such as mail service, billing, etc. are not a concern of this study although the study may, and often does, identify other weak spots in the corporation that could benefit from a QMS study.

Workflow—change of state—The next step in the development of the QMS is to define how the input changes into the output or product of the department. It is obvious that a series of transformations take place as tasks are performed in a department. There are some major steps signaling a change in state of the input as it becomes output. (See Figure 8.4.)

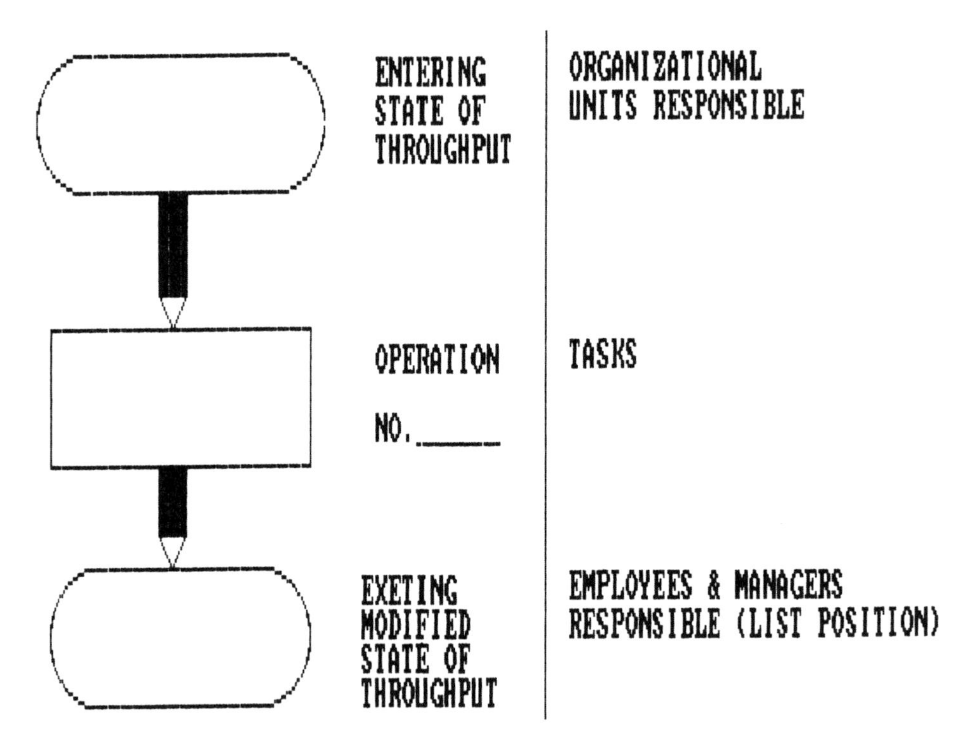

Figure 8.4 Workflow worksheet.

For instance, in a case reported for a communications area which receives foreign and domestic wire transmissions with instructions to transfer funds, the input is in the form of the incoming message.[2] In processing this message, many steps are taken resulting in four distinct intermediary and final stages (see Table 8.1).

The finished product is a correctly tested, translated message. It took many steps within four intermediary changes of state to result in each step's output.

Group Dynamics

The steps outlined above are the basis for the QMS. They are the systems portion of the task. These steps, as described below, are undertaken by a junior management team and other key people in the

TABLE 8.1 Example of Cable Unit Operation

I1	Incoming Message	
U1	Receive, Control and Translate	Control message by logging in and assure that it is properly translated into English text if needed.
U2	Routing and Investigation	Send to proper parties, set priorities and resolve problems.
U3	Unpack, Decode and Test	Convert to open language and check the validity of the instructions.
U4	Deliver and Record	Make sure that the message gets to the proper party and establish an audit trail.

department. By defining the system in the manner described, it has been found that the members of the team learn a great deal about their own operation, not only as far as it effects them but also how it effects others within the department.

Having completed the system analysis, they are now in a position to use their own data to develop all potential problem areas (deviations from normal) and to select the key quality deviations which need measurement and resolution. The method used for this process is the Nominal Group Technique. (See Figure 8.5.)

The Nominal Group Technique

The Nominal Group Technique (NGT) was developed by Delbecq, Van de Ven, and Gustafson around 1975 [3]. It is a form of structured brainstorming that has broad applications in quality control and other areas. As described by Professor Adam:

> The nominal group technique takes its name from the fact that it is a carefully designed, structured, group process which involves some participants in some activities as independent individuals, rather than the usual interactive mode of conventional groups. [4]

In other words, it is a form of structured brainstorming where all participants get an equal opportunity to contribute. Normal brain-

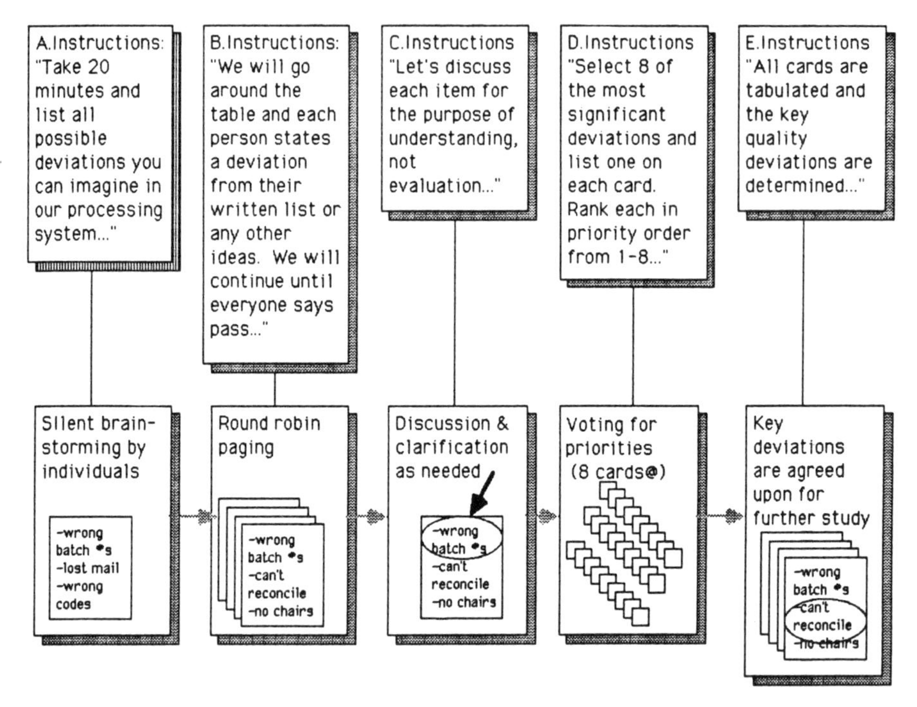

Figure 8.5 Steps in nominal group technique for developing key quality deviations.

storming often has the drawback that one or a few people dominate the activity.

The NGT method has been successfully used in situations where the group consisted of a wide spectrum of management levels. Experience with up to 7 levels of management shows that every level contributes equally well with no pulling of "rank." This requires a strong coordinator as will be discussed below.

Silent generation—The first step in the method is known as "silent generation." The name comes from the fact that the group is asked to list all possible deviations on one or more pieces of paper. By using the system boundary and the workflow organization, it is possible for the group to visualize all possible problems that can occur, likely or not.

By listing the possible problems in writing on a piece of paper and identifying the state from input to output where this deviation can occur, the participants are encouraged to think about the real and potential problems that can occur in their operation. This is often the first opportunity that they have had to reflect on the problems of the operation which result in bad quality.

Round robin paging—At the end of 20 minutes, the coordinator stops the writing and begins the round robin paging. This involves starting with one of the participants being asked to give one deviation and the area where it occurred. A person can state a deviation from his written list, or he can give a deviation which just occurred to him (perhaps triggered by a deviation someone else gave), or he can pass

The coordinator continues in turn for each member of the group. As the deviations are presented, they are listed on a flip chart. When every member of the group has had a turn, the coordinator goes back to the first person and continues the round robin paging until all members of the group have passed twice in a row.

Even though a member of the group has passed, he is asked again when his turn comes around once more. Some of the most important deviations have been generated by someone who passed several times only to produce a gem of an idea missed by all the others.

Clarification—As the flip charts become full, the coordinator removes them and posts them on the wall. A large room is needed for an NGT.

After all members of the group have passed twice in a row, the round robin is finished. The clarification stage is now required. Why clarification? Because the language we use in business is often ambiguous. Technical terms aside, common terminology has different meanings for different people. To "prepare work" can mean anything from receiving items to batching them to performing preliminary edits. It is important that all members of the group clearly understand each deviation.

When the clarification phase is in progress, people will ask for definitions of terms. It is not necessary for the originator of the deviation to explain the terms of his deviation. Any member of the group can do so.

Discussion, however, is limited to the definition. The validity or importance of the deviation should not be debated at this point. The next process of voting will take care of weeding out the important items from the unimportant. The coordinator must make sure that no politicking for specific points takes place.

Voting—The number of deviations listed is often very large. Between 70 and 120 deviations are often noted. It is apparent that not all of these deviations can be measured, nor that they are worthy of measure. To cull the most significant deviations, the key quality deviations, a process of voting is used. By voting, the group as a group selects the most important deviations. This form of consensus is important in getting the groups' co-operation in measurement.

The process is simple. Each member of the group receives eight card, three by five, preferably ruled or pre-printed. The members of the group are asked to select the eight most important deviations from those posted on the flip charts; putting one deviation on each card.

To assist them, it is well to advise the members of the group to extract from each flip chart one or two key deviations and list these. The selection of the final eight from the intermediary list becomes much simpler and the process is speeded up appreciably. It also appears that by breaking the task into components the group tends to have more similar lists than when this suggestion is not made.

When each group member has listed the eight most important deviations, one on each card, they are then asked to spread the eight cards in front of them and to select the most important deviation from those eight. This is marked on the card with a number eight and the card is turned over.

The group is then asked to look at the remaining seven cards and select the *least* significant deviation, marking it with a number "1."

The process continues for the remaining six cards alternating the selection from the most important to the least important and back. The last card needs no decision, it is of rank number "4."

The coordinator collects the cards and uses the results to determine the key quality deviations. Some judgement may be required here. The normal practice is to include any card marked with an 8 and any deviation that has a substantial number voting for it or generates a very high score.

Developing Measures

The fact that a deviation has been identified does not help by itself. It is necessary to get at the amount of the problem and, if possible, at the contributing source. A deviation might be that it takes too long to process a transaction. What is "too long?" How long does it take to process. Some transactions process faster than others. Why? What is the distribution of process rates and is it related to some factor.

Continuing this line of thought, what is the process average transaction speed? How much variability is there? What can the system do? What influences the speed of processing? Questions such as this lead to a better understanding of the key quality deviations. The measures which are needed are those that will answer the questions. When someone says that everything is under control, there is an inclination to ask, "How do you know?"

Use of NGT—The measures are obtained in a group session similar to the one which produced the key quality deviations. The main difference is that the voting step is not necessary.What is desired is as complete a set of possible measures of the effect as can be obtained. The step following the listing of measures refines the process to eliminate redundancy and impossibilities.

Problems and solutions—It is important that the group take the measures that they developed and prepare a form on which to record the measures as well as decide the "who," "what," "when," "where" and "how." This is the process of planning the data gathering. The "why" was covered in the previous NGT.

The results of the group's decision is documented and becomes a part of the tangible output of the QMS. The document is a plan which give the details of how the data is to be gathered and how it is to be recorded and interpreted.

The data collection form is attached to their plan. It is often in the form of a check sheet to record specific occurrences together with enough facts that when unusual conditions are indicated, there is enough information available to trace the problem. In the absence of this, another study is required to get the facts needed for decision making.

Measures which merely indicate that there are unusual conditions without allowing some action to be taken are not very useful. It should be the aim of the group to think through all actual and potential problems and to design the data gathering form to cover all foreseeable contingencies.

Once the data gathering method has been implemented, it is normal for the group to reconvene at periodic intervals to discuss the results and decide on further action. In order to avoid misinterpretation, these meetings should make use of a qualified statistician who understands the quality control process. Much of the benefit of QMS comes from such meetings. Unless managerial follow-up takes place, no results will come from the effort of the QMS.

Productivity Measures

Adam *et al.* use the measures differently. They are looking for productivity measures and so describe how to cast the results of the NGT on measures into productivity units.

As described above, productivity is a ratio of output divided by input. Consequently, Adam *et al.* divide the measures from the NGT into those pertaining to the output and those pertaining to the input. For instance, in dealing with computer downtime in one department, a quality measure consisted of the number of hours the computer was unavailable in relation to the number of hours the computer was required. Thus, if in a 35 hour week for two shifts, the computer was needed 70 hours per week but only up for 68.5 hours, the quality measure was 1.5 hours downtime divided by 70 hours required or 2.14% downtime.

Adam *et al.* would consider the 2.14% downtime as part of the output. In fact they have a reason to recommend that rather than use the concept of downtime, one use the positive concept of uptime. In this case they would say that there is 100% − 2.14% = 97.86% uptime.

They would then search for the input portion of the equation. In the situation they might find that the day shift has 25 people while the night shift has 15 people. At 35 hours/week this amounts to 1,400 hours. However, when the computer is down, the result is the need to catch up on the work not processed during the downtime. This may result in an amount of overtime. Suppose the amount measured at 45 hours, Adam *et al.* suggest that the input be both the hours of normal production plus the hours of extra work due to downtime. They would consider the input to be 1,400 plus 45 hours and divide this amount into the output of 97.86%. The resulting quality productivity measure 0.0677 would be reported as the productivity measure.

As things improve, the numbers increase. In fact, if everything were perfect (100% uptime and no extra work) the measure would be

0.0714 (100/1400). Unfortunately, such a measure has very little relevance to a department manager. The manager understands downtime and understands overtime but has trouble with the ratio estimator of the two.

It has been found that by working with quality measures alone rather than the productivity ratios, managers have relevant data with which to control their operations. For that reason, the QMS concentrates on proper planning for quality measurement rather than worrying about productivity measures.

Implementing a QMS

The mechanics of the Quality Measuring System were described above. In addition to the mechanics there are some important executive considerations that need attention.

Executive Management Commitment

General Eisenhower once explained the concept of leadership by comparing it to moving a string: it is easier to pull than to push. This is relevant to a basic requirement for quality programs. Programs which are pushed by executives onto their subordinates will fail while those programs which are lead by the executives active participation will succeed. The responsibility for quality rests with executive management and cannot be delegated any more than financial responsibilities.

Need for commitment—It is hard to conceive of a corporate president who appoints a comptroller and instructs him, "Just send me a report once a month or quarter telling me that everything is alright. Look into how we should budget and do accounting and then set everything up accordingly."

It might be conceivable that a president do such a foolish thing if it were not for the fact that the owners, stockholders and agencies such as the IRS and SEC required the president and the board of directors to be accountable for their financial skills. There is no similar driving force with regard to quality. The only thing coming close is the consumer in the market place.

If one considers all trade, domestic as well as international, the record is clear that the president who delegates his responsibility for quality will lose out in competition to the president who is active in managing quality. The automobile industry, the high tech industries and many others have found out too late that their poor performance can lose markets. They then call for government intervention which, if given, means that the consumer underwrites their ineffectiveness.

Even managers considered outstanding by those who see their financial and marketing expertise cannot meet the competition. In a recent article stating that the government would not insist on a continuation of Japan's "voluntary" export quota of cars, Chrysler Chairman Lee Iacocca is quoted as saying: "This is a sad day for America — for American workers and American jobs." He is admitting that the greatest industrial nation cannot compete in the automobile business without government protection [5].

It should be noted that neither GM nor Ford has been reported as concerned. That might not be an accident. The chairmen of both of these manufacturers are actively pursuing the quest for quality. Most likely, they feel that they can compete.

Until bank deregulation, bankers were protected. With a little bit of care, it was hard to go wrong. Not any more. Deregulation has created a competitive market which more and more bankers are finding are impinging on practices which were previously acceptable [6]. For this reason, bankers are becoming interested in starting quality control programs. They should be aware that these programs will not work unless the bank executives work at the process of quality as hard as they do on asset/liability balances and profit plans. Delegation just will not do. The successful bank executive has recognized the situation and is as active in the managing of quality as of finances.

Nature of commitment—Obviously, executives are busy directing the work of others. One of the areas that they need to address is the obstacles that their subordinates face in carrying out their duties. Whether this is lack of resources or lack of adequate personnel makes little difference. The executive needs to know three things:

1. Are there obstacles to perfect performance (it is assumed that the executive has adequately defined what constitutes performance).

2. If there are obstacles, what are they; if not, how do you know that they are not hidden.
3. What needs to be done to remove the obstacles to achieve good performance.

The two implications are that a statistically sound tracking system exists and that the executive takes personal charge. Executives are well advised to schedule a portion of their time to see the workers and talk to them. This discussion should be about the job, what could make it easier, what does it consist of. If the workers cannot explain their job in a satisfactory manner to the executive, they probably don't know it fully.

If top executives are known for their surprise visits, it will filter down the line. No manager likes surprises. Unpleasant surprises usually happen because the manager has isolated himself from the reality of the workplace. Often, top executives are isolated. They can break this cycle by careful planning of their time to include the topic of quality.

It is unsatisfactory to use breakfast, luncheon or dinner meetings to find out what is going on. The dining room is not the workplace and tends to frighten the workers that could "tell it as it is." Why is the turnover so high? Talk to the workers.

Executive involvement requires a balanced judgement of time. A great deal of time may be needed just to gain the confidence of the employees if this area has been neglected.

If the method of QMS is used to develop the measures, it is important that the executive schedule enough time to get a thorough briefing of the subject, determine the participants and give his blessing by writing a personal memorandum to each participant outlining what he expects and detailing the meeting dates and times. Such a memorandum gives the sessions the significance they need to get started. The executive should also commit himself to the schedule and allow no interference with it.

Selecting groups for participation—The people who participate in the QMS should be selected with great care. They should be in a position to participate fully. A person who will be on vacation during any of the sessions cannot be a proper participant. Missing a single session puts the person at a great disadvantage compared to the others.

The first person to select is the coordinator. This is the facilitator of the group. It is the coordinator's task to instruct the group in what has to be done, help them over rough spots and collate and publish the results.

The coordinator should be a person skilled in the task. An amateur can ruin the process, ending in more frustration than results. The coordinator can be from within the organization or from outside. In the beginning, a skilled consultant is a good investment to train internal coordinators in the task.

The coordinator should be someone who can garner sufficient respect for his role that the meetings proceed smoothly. Professional facilitators or teachers make excellent coordinators once trained in the method. The coordinator should not be an inhibitor. If the coordinator is not trusted, the results can be greatly biased. Sometimes the bias can make the whole system fail. If the professionalism of the coordinator is recognized by the group, the coordinator will succeed.

The coordinator's first step is to be familiar with the language used by the organization doing the QMS. A good way is for the coordinator to spend time in the area and to prepare a first draft of the system boundary. When the coordinator feels that he understands the workflow enough to sketch it out, the coordinator has learned enough to proceed with the first meeting.

The executive selects an upper management group of perhaps three or four people. This group is continually updated by the coordinator and advises the coordinator of any problem areas such as major deviations not found by the working group. The upper management group acts as a consulting group to make sure that the QMS will succeed.

The use of the upper management group has a second purpose. It keeps the upper managers involved in the process and prevents a feeling that the results are not their responsibility. By reviewing what is going on, they have a significant input to the process.

The working group consists of 12 to 15 people involved in the process. They will normally be first and second line managers. Some key employees, staff persons or customer representatives may also be present. This group does the actual work and so becomes the "owner" of the product. As a result, interest in the process is high and later cooperation is almost always assured. In effect, they are being given a powerful tool with which to handle some complex problems of their operation.

The experience has always been favorable. The working group enjoys its prestige and willingly contributes significantly to the organization. They are also in the best position to effect the small changes that can clear up many of the minor deviations that were left for later analysis.

Once the groups are determined by the executive, four meeting dates are set. The four meetings last from three to four hours each. Normally they are spread over a period of three weeks although it is possible to do the whole session in a week.

The executive drafts a memorandum to the participants outlining the dates and asking them to attend. Of course, copies of the memorandum with an explanation of the task go to the participants' supervisors if they are not otherwise involved in the upper management group.

First Meeting

The first meeting is attended by everyone. The executive opens the meeting with a short reiteration of the contents of the memorandum. He introduces the coordinator and turns the meeting over to him.

The coordinator reviews the whole concept and then describes the system boundary in detail. The draft system boundary for the group is mounted on a flip chart. It is purposely kept incomplete. This gives incentive for the group to correct and add to the draft. The coordinator draws in the changes on the chart. This approach serves two purposes:

1. To get a correct version of the system boundary
2. To act as an icebreaker

When the system boundary has been completed, a short break is indicated. At this time, by prearrangement, the executive and the upper management group withdraws from the room. Henceforth, they operate in the background as described above.

On reconvening, the coordinator explains the workflow. He gives a neutral example. The neutral example allows for an understanding of what is needed without acting to influence the group.

The group is divided into four or five smaller units of three or four members each. These units were preselected in such a way that every group has someone from each major area of the operation. In this way, each group can develop the whole workflow for the organization. Those not previously familiar with all phases of the operation learn from each other. This learning experience alone is valuable.

The coordinator circulates among the groups and gives guidance were required. A careful balance must be maintained between helping a group that is stuck and influencing the result. The coordinator wants to do the former not the later.

When the groups have completed their worksheets, the coordinator collects these. The group is finished for the first session.

The coordinator reviews all worksheets and creates an amalgam from the result. Normally, it takes three to four changes of state to go from input to output. The coordinator summarizes the result and has the system boundary, the coordinated workflow sheets and the summary typed in preparation for the next session.

The coordinator meets with the upper management group to go over the result and make any changes that might be indicated. Changes are rare when a skilled coordinator runs the session.

Second Meeting

The second meeting is with the working group alone. The purpose of the meeting is to develop the key quality deviations using the NGT method with voting. To open the meeting, the coordinator distributes the package of material containing the system boundary and workflow. This is reviewed for correctness. Any adjustments are noted. Then the NGT begins.

The NGT requires the use of flip charts and a room large enough so that the flip charts can be mounted and easily viewed by all members of the group. It can be expected that between 10 and 15 flip charts will be posted on the walls. One company arranged a long wall to be used by installing a piece of molding containing enough spring clips to hang all the flip charts. Others use a blank wall and drafting masking tape. The drafting masking tape is able to hold a flip chart but does not leave any marks when removed.

For the voting process, 8 three by five inch cards are needed for each member of the group. These cards should be lined rather than blank. The top of the card should be pre-printed if possible with headings of "Dev #," "Deviation" and "Vote" (see Figure 8.6).

```
Dev. #                Deviation                      Vote
--------  ------------------------------------  ┌--------┐
                                                |        |
.......   ..................................    |        |
                                                └--------┘
.........................................................

.........................................................

.........................................................

.........................................................
```

Figure 8.6 NGT voting card.

The use of a pre-printed card avoids the confusion between deviation number and vote. The coordinator sequentially numbered each deviation as it was listed on the flip chart. It has been found that both deviation number as well as the description of the deviation are needed. Often the description is hard to decipher. Sometimes it is different from what appears on the flip chart.

Third Meeting

The third meeting uses the same facilities as the second meeting. The coordinator begins the meeting by announcing the results of the group's vote, a copy of which is given to every member of the group. After a brief explanation of the procedure and description of what constitutes a measure the group again proceeds with an NGT. The session ends when the clarification has been completed.

The most important aspect of this session is to make sure that the group members have a clear understanding of what constitutes a measure. Up to this point they have been considering what can go wrong in the process. The concept of how to measure the effect is not easy. In some cases direct observation and recording is possible such as the number of typos in a letter. In other cases, indirect effects must be used to trace back to a result such as the number of customer inquiries related to a specific transaction.

If the group understands the nature of measurement they will find the session easy and productive.

Fourth Meeting

The last formal session of the group for QMS is used to develop the actual check sheets with which to record the data and develop a plan of how these data are to be used.

The coordinator begins the session with a summary of the measures that were developed in the previous session. He then reviews various forms of control charts. This is done to give the group a perspective of what can be done with the data.

The group then proceeds to develop the recording form for the various measurements. Some constants on recording forms or check sheets are the date, the person recording the data, the data itself and observations relating to the data. The exact nature of each of the principle parts of the form is developed by the group.

The group will also develop a plan on how to collect the data, who will be responsible for the collection process, how often to collect the data, to whom the reports will be given and what is to be done with the data.

When the group has developed the completed plan it has done the task it set out to do. The product of the effort is the system boundary, the workflow analysis, the list of all deviations, the list of key deviations, the list of measures, the form for recording the data to be measured and a plan for implementing the measures.

Executive Management Approval

The results of the working group's efforts are reviewed by the upper management group and the executive who commissioned the task. In this review the coordinator goes over all that is new to the group and presents the plan of action. The group can accept the plan of action or modify it.

If the plan is accepted or the modification small, it should be put into effect at once. Acceptance is interpreted to mean executive management's commitment to monitor the process and to help it to succeed.

The most important aspect is immediate feedback while the members of the group are still in a very positive frame of mind to implement the system. Any delay of even a week or two has been known to setback the operation and make it less effective than it could have been.

If the plan is not acceptable, it is important that the management group meet with the working group to explain the reason why the plan falls short of acceptability. The working group may be able to modify it to meet the management criterion. At least they will have feedback which is of importance if they are ever again called upon to act as a group.

Results of Implementation

A number of examples have appeared in the literature illustrating how the method has been applied [7].

Some Examples

As a result of using the QMS some glaring operational faults have been discovered. The problems existed for some time but no one dared to bring it to management's attention.

For instance, a company operating in Mexico had large turnover of personnel. The company had excellent operating conditions as well as progressive management. The workplace was clean, well lit and specially designed for the paperwork task in which the company was engaged. The company operated on a paternal basis giving employees proper training, recreational facilities and even providing bus transportation to and from the employees' district because there was no public transportation available. Salaries were very good in comparison with other firms in Mexico.

The QMS uncovered two problems which had not been known before. One of these was that overtime pay was not given for two weeks because of an operating procedure in the payroll department. The workers did not understand why they did not get what they thought they had earned in the week they earned it.

The second problem related to the bus system. Through the QMS it was discovered that the rented buses were not following the prescribed routes. What is more, it turned out that the drivers were picking up non-employees for pay and that often there was not enough room on the bus to bring the employees to work.

The first problem was easily corrected with a change in the payroll system. The second problem was corrected in two steps. The first was to get employee representatives and management to review the bus routes to chart the best for the employees. The second step was to work with the owner of the busses to correct the abuse of his drivers and to monitor the system's operation.

The results of these actions by management on the basis of QMS data was to cut the turnover from its initial high 85% to about 20%. Other firms in the area are still plagued with the higher turnover rate.

While the two examples illustrate a small result, the measurement in the same plant disclosed a surprising source of error. The process required a step in which an account number look-up was needed for batches of work. The number was then transcribed by the look-up clerk onto the top item of the batch. The batch data was then key entered. Entering the wrong account number caused all subsequent work to be incorrect, something that would not be discovered until a customer complaint was received.

In addition to customer dissatisfaction, the misposting meant lost funds availability of substantial amounts. It was thought that the problem, a key quality deviation, was misnumbering the accounts or, more likely, misposting by the key operator. Report forms were established for both the numbering operation and the key operation. In both cases samples were used to check the quality of the work. On the report form was noted the correct number as well as the wrong number. The reason was a suspicion that number inversion was the culprit.

As it turned out, the problem was in the numbering operation. It was not a matter of inversion of digits. By looking at the correct number versus the incorrect number it became apparent that certain digits were often involved in the mistake. On investigation it turned out that the operators had been taught to write numbers in a fancy script. This is fine if the number is written carefully but becomes a problem when it is written hastily. A seven could look like a four or a nine. The number "two" was indistinguishable from the number "three" or "five."

The solution was to train the entire plant in writing a standard set of numbers without frills. Wherever numbers were used in the operation, signs were posted with the standard numbers shown. New employees were trained in the standard way of writing numbers. The problem of misposting virtually disappeared. Cash flow improved. People were happier because they were no longer blamed for problems which they thought were not of their doing.

QMS Does Not Stand Alone

As can be seen from the few examples described above, the QMS is a beginning and not an end unto itself. The fact that measurements have been initiated and even the fact that they disclose problems and give a clue to the source of the problems does not solve the problem. What is required is a series of management tools and management action in the use of these tools. The tools are control charts. The data from the record forms does not show much by itself. It requires interpretation that distinguishes between operator controllable and management controllable sources of the problem.

Only management can change global questions such as the three mentioned above. It takes a perceptive management to solve the problem. In the case referred to above, the management was following Dr. Deming's 14 points described in Chapter 9.

QMS followed by tools and action are the key to the successful implementation of a quality management system.

Summary of QMS

The use of the Quality Measuring System (QMS) to initiate a quality management program in a bank or any industry has given good results wherever tried. The method has great benefits of both a direct and indirect nature. Portions of the method have application in areas other than quality control.

Advantages

The direct advantage is the structured method that gets at the heart of what to measure in as fast a way as possible. When properly used under the guidance of a skilled person supported by a participative executive management, a plan for implementation can be formed in one to three weeks.The plan can be carried out immediately and measures with possibly some results can be expected in a few weeks or months after implementation. The time depends on the type of measures and the circumstances of the operation. If the operation is "clean," that is to say with all jobs well defined with no redundancy of tasks, then the time needed will probably be short. Unfortunately, such an operation is rare.

Another direct advantage is that the method predisposes the first and second line management team towards implementation and use of the results. The whole measuring system is theirs, they have "bought in."

An indirect benefit is the formalization of just what the department is responsible for: what products are their responsibility and which are not. Junior managers have the opportunity to get an overview of the whole job including what takes place in areas outside their immediate jurisdiction. Further they learn a method of group problem solving and structured brainstorming. In a number of instances, supervisors applied the same technique in their own area with great success.

Application

The method of QMS has been successfully used in planning the implementation of major systems, in risk analysis, in service applications and in manufacturing.

A particularly powerful part of the process is the systems analysis. A similar approach has been used by IBM called Department Activity Analysis (DAA) [8]. Such tools are all good as long it is recalled that they merely start a process of analysis and are not an end in themselves. Like all tools, there is still the need for a skilled professional manager who knows which to use where and what to do with the results.

References

1. E. E. Adam, Jr., J. C. Hershauer, W. A. Ruch, *Measuring the Quality Dimension of Service Productivity*, National Science Foundation Grant No. APR 76-07140, January 1978.
2. W. J. Latzko, "Quality Productivity Measures — Participative Management," in *Thirty-Fifth Annual Quality Congress Transactions* (San Francisco: American Society for Quality Control, 1981), p. 392.
3. A. L. Delbecq, A. H. Van de Ven, and D. H. Gustafson, *Group Techniques for Problem Planning: A Guide to Nominal Group and Delphi Processes*, Glenview, IL: Scott, Foresman & Company, 1975.
4. Morris, W. T. and The Ohio State University Productivity Research Group, "Measuring and Improving the Productivity of Administrative Computing and Information Services: A Manual of Structured, Participative Methods," National Science Foundation Grant APR 75-20561, p. 40.
5. DeMott, J. S. "Stop Sign," *Time* (New York), March 11, 1985, p. 40.
6. Latzko, W. J. "Quality Will Determine the Winners in Competitive Banking Market," *Banking Systems & Equipment*, Vol. 21, No. 5, May 1984, p. 204.
7. From William J. Latzko, "Dr. Deming's 14 Points in the Service Industry," A speech presented at the Annual Conference of the Administrative Applications Division of the American Society for Quality Control, Williamsburg, Va., March 21, 1985.
8. Melan, E. H. "Process Management in Service and Administrative Operations," *Quality Progress*, Vol. 18, No. 6, June 1985, pp. 52–59.

9

Human Resources in Bank Quality Control

Banks have always realized the importance of people in their operations. Without a properly trained staff the functions of banking would be a shambles. A major problem that banks face today is the rapid advance of technology, of government regulations and of massive competition. The result has been that banks have changed their way of doing business and in turn this resulted in what workers perceive as the dehumanizing of the workplace.

Workers that are improperly motivated, improperly trained, and poorly supervised tend to make mistakes. Much emphasis has been placed on the motivational aspects of the workers job. Too much perhaps. A properly motivated worker without tools and poor supervision will just become frustrated.

Nevertheless, there has been a large movement in banks toward participative management. It was felt that if workers had a larger say in their job they could solve the problems facing management and morale, quality and productivity would all improve. Most of the effort made was in the direction of quality circles. Bank officials would proudly state how many quality circles they formed. Their measure of success was the number of quality circles formed.

Quality circles certainly have a place in the operation. If properly used they can be extremely effective. They are not a substitute for a program of managing quality. Unfortunately, quality circles are often substituted for such a program of quality management with devastating effect. In the end everyone is unhappy, workers and management. In this instance, banks are merely following the fashionable trend of other American industry.

Since participative management and quality circles do have a significant role to play in managing quality, it is important to examine the topic, its background and how to obtain the desired results from such methods.

Participative Management

In general, participative management does not imply participation in policy making but implies participation in solving the problems of the workplace. Again, these are more often the problems perceived by management than those perceived by the workers. Even when there is an overlap, there may be no agreement as to the solution.

Dr. Juran distinguishes between three levels of participation (see Figure 9.1.):

1. Defense against unwarranted blame. Operators defend themselves against such blame by pointing out deficiencies in the system of self-control; i.e. management has not met the criteria. This feedback is forthcoming even if management displays no interest in operator feedback generally.
2. Collaboration on operator-controllable defects. The operator's responsibility with respect to these defects is so clear that there is seldom a problem of securing feedback from the operators. While some of this is defensive, operators generally welcome being consulted as to their observations, experience, theories, etc.
3. Collaboration on management-controllable defects. This level of collaboration requires special provision such as

THREE LEVELS OF PARTICIPATION BY WORKERS
Participative management implies participation in problem solving, not necessarily in policy formation.

Level One
Defense against unwarranted blame.

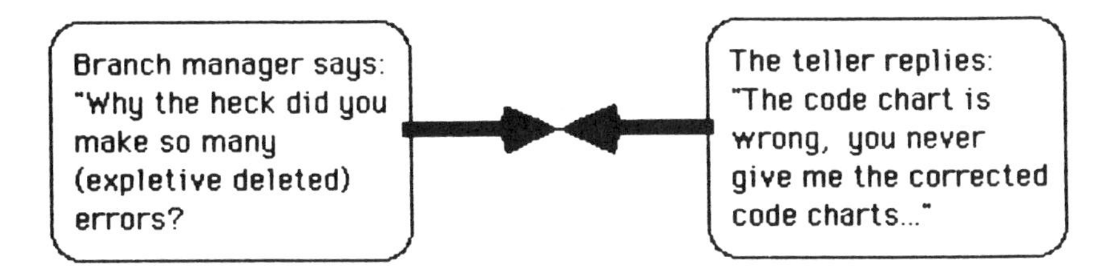

Level Two
Collaboration on operator controllable defects.

Level Three
Collaboration on management controllable defects.

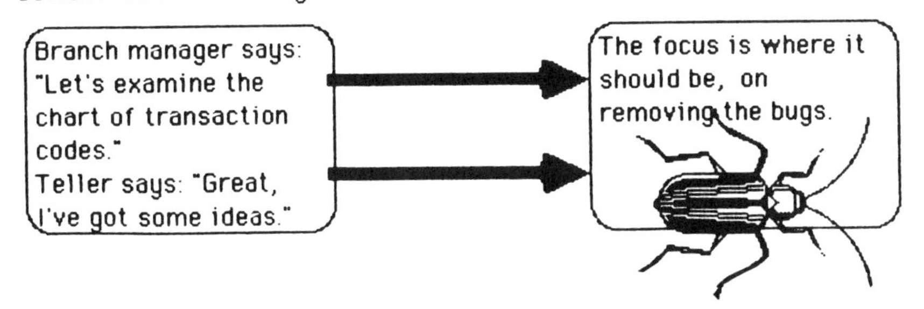

Figure 9.1 Three levels of participation by workers.

> suggestion systems, Labor-Management Committees, QC Circles, etc., to legitimize the activity. [1]

Dr. Juran's sophistication is not shared by most bank executives. They are frankly puzzled by the distinction between management and operator controllable defects. Don't the operators make the mistake? How can management control these except by punishing operators for mistakes? This common perception is the reason why so many executives believe that quality control circles will solve their problems. The ones who make the mistakes should know the most about how to stop making them.

To address this topic it becomes necessary to look at modern concepts of management, concepts which are not obvious but which have a proven track record.

Management's Responsibility

The easiest yet the most erroneous view is that the one who makes the mistake is at fault. This simplistic view of operations in the workplace should not require examination but, unfortunately, it so widely held that it needs to be addressed.

Many authorities have stated that 85% of the defects are management controllable, 15% worker controllable [2]. How can this be?

Management has the responsibility to set policies, corporate objectives (including the constancy of purpose principle), planning how to carry out the tasks, organizing the task, getting the tools for the job, training the workforce in how to do the job, make the parts interact properly, determine that the job is done properly and remove obstacles to the proper performance of the job. This is no mean feat.

Management's failure anywhere along the line in the tasks set out above will cause defects. These defects are delivered by the worker, but are part of the production system over which management has sole control and responsibility. Supplying workers with partially legible forms, incomplete procedures, poor computer systems, inadequate working conditions, poor training (or no training at all), poor supervision, among many others will cause defects over which the operator has no control. To blame the operator for such defects is not only foolish and stupid but dishonest.

All of this is easy to see in the obvious cases. Unfortunately, there are many instances where it is not easy to see whether a defect is management or worker controllable by merely looking at defects. There is a way of making this distinction. The method is Statistical Quality Control (SQC). This method will determine if the cause is systemic (management controllable) or special (operator controllable). Operators and supervisors equipped with and trained in the use of SQC tools can intelligently participate in the solution of problems. Those not so equipped will continue on witch hunts and finger pointing exercises.

In addition to these tools, a management philosophy must pervade the organization to make effective use of these tools. This philosophy is most clearly stated by Dr. W. Edwards Deming, the man whom the Japanese credit with their remarkable postwar recovery. Dr. Deming taught the Japanese not only the technical aspects of SQC but the management tools embodied in his 14 points for management. Many companies in the United States as well as Japan have found that the 14 points work.

Dr. Deming's 14 Points

Dr. Deming's 14 points are the result of his observation of ingredients that make a firm successful. The companies that employ the philosophy have been uniformly successful. Many of them have made remarkable turnarounds from previous disasters.

The 14 points are related and make a synergistic group. One cannot pick a point here, another there and expect those points by themselves to give solid results. For instance, one point is not to buy on price tag alone. The purchasing department would be hard put to do this if barriers exist between departments which would cause the purchasing department to try to optimize its position regardless of the impact on other departments.

The following is a short description of the points following them as outlined in Dr. Deming's book, *Quality, Productivity, and Competitive Position* [3].

1. Create constancy of purpose for improvement of product and services—This means that in order to stay viable as a business, management must pay attention not only to the problems of today but also to the problems of the future. What products can be sold in tomorrow's market? What is acceptable to our customers? How can we improve current products? To accomplish this requires investment in research, innovation and people. Equipment must be kept current to give employees the tools needed to be competitive.

2. Adopt the new philosophy. There is no need to tolerate error—If management recognizes that they are the major controlling force to reduce mistakes and take appropriate action, the error rate will diminish. Supervisors must be trained to look for the small problems, those that in themselves appear not to be significant, are accepted as a way of doing business, yet cause error and lost productivity.

3. Cease dependance on mass inspection—As explained from page 62 on, 100% inspection is much less accurate than most people believe it to be. Only 60% to 80% of the errors are caught. Computer verification is not much better for a number of reasons. This type of inspection is not only expensive and ineffective but gives a false sense of security to management and supervision. It can be shown that the same amount of money spent on improving the process will give far greater payback than mass inspection could ever yield.

4. End the practice of awarding business on price tag alone—Most bank purchasing agents pride themselves on the money saved by buying forms on competitive bids. The money lost in computer reruns needed because of the poor forms, the lost production because of second copy illegibility are never known to these purchasers. Besides, they are not rated by their bosses on how good the material works in the process but by how little it cost; as long as the performance is marginally good.

Saving 10% on an encoder ribbon that breaks continually causing encoder production to drop from 1,200 items to 800 or less is a concern of the operating department, not purchasing.

5. Constantly and forever improve the system of production and services—This is the aim of many managers. They cannot do it. Why? Because they do not truly know what makes their system tick. Without guidelines that can separate the common causes from the special causes, they cannot tell if their system is in control or not. If they try to alter a system which is not in control, they are guessing as to the cause of problems and have more than a 50-50 chance to make the system worse. Improvement requires knowledge. This is knowledge that can only be obtained with some basic tools, the tools of statistical quality control. Anything else is a guess.

6. *Institute modern methods of training on the job*—Much training is by experience—usually bad—on the job. People often do not know what their job is. Little is written about the task, procedures are outdated. Supervision is not consistent and thus a poor guide. Besides, they do not know what their job is either. Defining the job and training the operator or lending officer in the what they are supposed to do and what they are not supposed to do is key in reducing errors.

When has an operator or loan officer learned the job? Again, there are tools that can clearly define when the job has been learned and no further training is needed. The same tools can be used to tell if a person will ever be able to cope with the job or not. What are these tools? Properly applied control charts.

7. *Institute modern methods of supervision*—A supervisor should not be merely a judge. A supervisor should be a coach. A supervisor or manager that does not know the job at hand is a hazard to the bank. Such a person can cause a formerly good operation to disintegrate practically overnight. Since there are many such supervisors and managers around, it is an indication that their boss does not know the job either and/or is doing it poorly. (See Figure 9.2.)

The supervisor's job is to remove the barriers that cause the worker not to be able to perform the job properly.

8. *Drive out fear*—Why do workers accept less than satisfactory tools and/or operating conditions? Why do they not tell their boss about such problems? Because they do not want to "rock the boat." Those that object to the way things are done in this department are troublemakers. Troublemakers do not last long in this organization.

A supervisor should
not be a judge,
but should be a
coach.

Figure 9.2

It is easier to do nothing than to do something. This is true for workers and managers. To do something requires a change from accepted patterns and people culturally resist change. It means taking a chance. If the action does not succeed, the change maker is faulted; if it does the change maker is often not rewarded because that, after all, was his job. It is better to do nothing; one does not get in trouble that way.

Fear is holding back innovation, improvement in quality and productivity. It is caused by management methods and can be removed only by management action.

9. Break down barriers between staff areas—Departments that do not cooperate with one another will suboptimize the bank. The purchasing department that is judged by their savings only is not interested in the problems of the operations department, especially if the problem cause is not readily apparent. Production goes down when the new ribbons are used but that is thought to be the fault of the operators. They should work harder.

It is only when all departments work for a common objective that many of the small problems can be resolved. Very little can be accomplished when each department is forced to suboptimize its efforts to stay in business. (See Figure 9.3.)

10. Eliminate numerical goals for the workforce—Slogans abound and are posted on walls. "Do the job right the first time" sounds just fine, the question is how can it be done right the first time when the input forms are defective? To tell a worker to increase productivity without giving the worker the tools or training necessary to accomplish this objective will only frustrate the worker. It could even make things worse. A worker in control, as shown by a control chart, cannot improve his error rate. Any attempt on his part to do so will result in making his work worse.

11. Eliminate work standards and numerical quotas—A major cause for error is the establishment of work standards or quotas. There are two problems with these. The first is the signal to the worker in spite of all protestations that numbers are more important than quality.

The second is that many goals and work standards bear no relationship to what can actually be performed. It should not be necessary to say that if they are based on averages—and they all are so

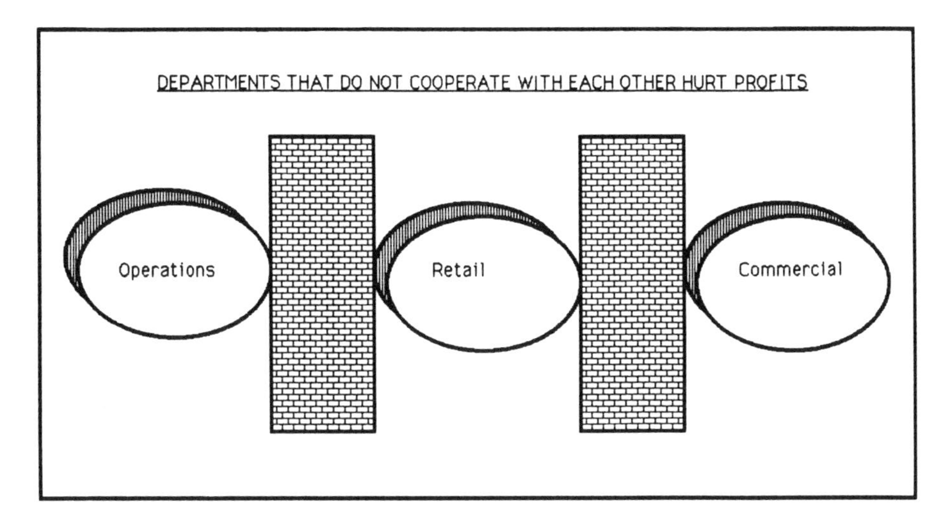

Figure 9.3 Break down barriers between departments.

based—that some of the values used to set the standards were well below the "average" and some well above. One thing is certain, smart workers will not exceed the quota if they can help it. They know that if they do, the quota is raised.

If the quotas are too high the result is to frustrate the workers. In neither case, quota too low or too high do the quotas work and in all cases they signal that numbers are more important than quality.

12. Remove barriers that hinder the hourly worker —And management as well. When workers, clerical, professional, management do not perform in accordance to expectations the question is can they? With a control chart that question can be answered. If the problem is under their control appropriate action can be taken as discussed elsewhere. If the problem is not under their control, they cannot take action themselves. Management must find the cause for their inability to perform as expected and remove the barriers that prevent them from doing the job satisfactorily. (See Figure 9.4.)

13. Institute a vigorous program of education and training—Work changes. New methods, new pieces of equipment and new products appear all the time. This is especially true if management follows

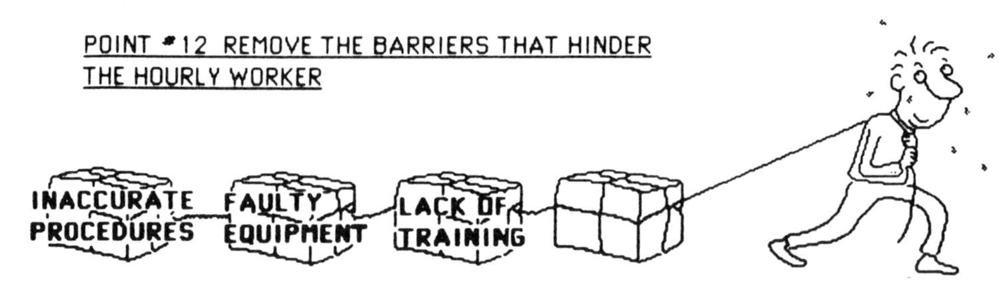

Figure 9.4 Remove barriers that hinder the hourly worker.

Point 1. closely. The new items change the way the task is being done. Workers need to be continuously trained in how to do the job correctly. This is best done by formal training by qualified personnel.

In addition, it pays the bank to provide general education to their employees. Some special skills are needed if the bank uses quality circles or the like. The requisite skills needed are not available in the workforce. In fact, the necessary educational background to understand the basic skills for quality circles are often lacking. The bank is well advised to make up for this deficiency.

14. Create a structure in top management that will push every day on the above 13 points—The management of quality comes about only by the sincere hands-on application of the philosophy described above. Without acceptance and use of the points throughout the organization, they will not work. And failure to follow these points will reduce the competitiveness of the organization. With the changes in banking, bankers need every advantage they can get. In 1984 and 1985, several bank failures and massive write-offs as well as penalties were recorded for major banks including Bank of America, Continental Illinois, First National of Boston, First National of Chicago, among others. Who will be the survivor? Those that are competitive and to be competitive banks need the 14 points.

The Japanese Experience

In an effort to learn the formula for success to Japanese competitiveness, a number of American firms sent people to study the Japanese method of quality control [4]. What they were often shown was the latest development in Japanese production techniques. At the time the latest developments were quality circles. The more recent methods of Toyota's Kanban and Professor Tagushi's experimental design and loss functions are just now filtering into American industry. Banks have yet to discover them.

The American researchers returned convinced that the secret to competing with the Japanese lay in the installation of quality circles. As a result it became modish to install quality circles. As Dr. Sandholm wrote, "New concepts and methods often create interest. Consultants always pop up, taking the opportunity to exploit this interest for their own commercial purposes" [5].

To better understand the use of quality circles, it is worthwhile to trace a little of their origins. In this development can be found the principles that must be considered to use this powerful tool correctly.

Background and Misconceptions

In 1950 the Japanese invited Dr. W. Edwards Deming to lecture on quality control. Dr. Deming not only lectured to engineers but also to top managers of some 60 Japanese firms. He taught the managers the essentials of the 14 points which started them on their road to capturing world markets. The Japanese honor Deming's work by their top award for quality, the "Deming Prize."

Dr. Juran's contribution—In 1954, Dr. Joseph M. Juran followed Dr. Deming and also taught management concepts to the Japanese. As part of his lectures, Dr. Juran taught his great concept of managerial breakthrough. This is a structured method for effecting change to a system in control. As outlined by Dr. Juran, the method uses a group of management personnel to focus on specific problems and resolve them.

However, quality circles are not, contrary to popular belief, an American idea. "Although American methods and the ideas of Dr[s]. Juran and [W.] Edwards Deming were most influential in Japan, there can be little doubt that circle activities reflect a creative synthesis of western ideas and Japanese organizational practices. Thus, the difficulty of transplanting back to American soil must not be underestimated" [6].

Cultural differences—Some of the differences that must be considered are the culture and the nature of the workplace. Japanese lifetime employment is well known. A worker hired by a Japanese company remains with the firm. The firm is paternalistic to the extent of providing the worker's wedding reception, providing athletic outlets, etc. In return, the worker has a strong loyalty to the firm.

The workplace concept is different as well. Japanese supervision and management spend long years as employees learning the ropes. Promotion is on the basis of longevity. Conformance and cooperation are prized attributes. The confrontational style of management common in the United States is not considered good form in Japan. People help each other to get the job done. They subordinate their efforts to that of the company. This is very natural since they can identify their welfare with that of the company.

Japanese workers are more independent than those in the United States. They have a greater control of the workplace and their work methods. They are not saddled with the Taylor concept that only management knows what is good for the workers, so the workers can park their brains at the door. The Japanese worker takes a healthy interest in his job.

It is common in Japan for workers to meet after work to socialize and discuss their company. Since their lives are so wrapped up in the company, the company's affairs are theirs as well. It should also be noted that the average Japanese worker has a higher level of education than the average American worker (12th grade as compared to 10th). In addition, Japanese high schools now teach quality control as part of their curriculum.

Dr. Ishikawa's contribution—Dr. Kaoru Ishikawa is credited with the invention of quality circles. Dr. Ishikawa converted Dr. Juran's managerial breakthrough technique into the Japanese idiom. Where Dr. Juran uses a management committee for the selection and solution of problems, Dr. Ishikawa uses the workforce. By their outlook and training, they are well qualified to solve problems.

In addition Dr. Ishikawa invented the now famous "cause and effect" diagram often called by his name. The cause and effect diagram is an organized way for developing the potential cause of a problem. It corresponds to the analysis for cause by the managerial breakthrough steering committee. This method is employed in almost all problem solving in Japan since it presents a compact pattern of relationships. This allows Japanese managers to see that their engineers have covered all cases.

Results of Japanese Experience

The Japanese use of quality circles has stirred a great interest in American management. As explained before, it seemed as though the whole success of the Japanese manufacturing system was due to these quality circles. Nothing could be further from the truth. Dr. Ishikawa himself stated that his measurements indicated that at best the quality circle movement in Japan accounted for no more than 10% of the improvement [7].

Professor Cole of the University of Michigan has made a study of quality circles and in a 1981 conference listed the following misconceptions about quality circles:

> One of the common misconceptions prevalent in American management circles is that QC circles are primarily responsible for Japan's remarkable success in improving product quality. Yet, the development of the circles came at the end of a long chain of activities designed to establish product quality as a major management priority. In particular, the spread of a quality consciousness throughout management and technical groups, together with quality audits and statistical problem solving techniques, were all well diffused before the initiation of QC circles. [8] (See Figure 9.5.)

Professor Cole's second point is that while government forces helped to encourage Japanese manufacturers to ship only top quality material (The Export Inspection Law of 1958) they had nothing to do with QC circles. It should also be noted that the Export Inspection Law came eight years after the quality initiative of private industry. In fact, the Deming prize has had more impact on improving Japanese quality than this law.

Thirdly, Professor Cole notes that QC circles are not widely diffused throughout the workforce in Japan. He says that they are principally a blue collar phenomenon and only recently introduced to white collar workers.

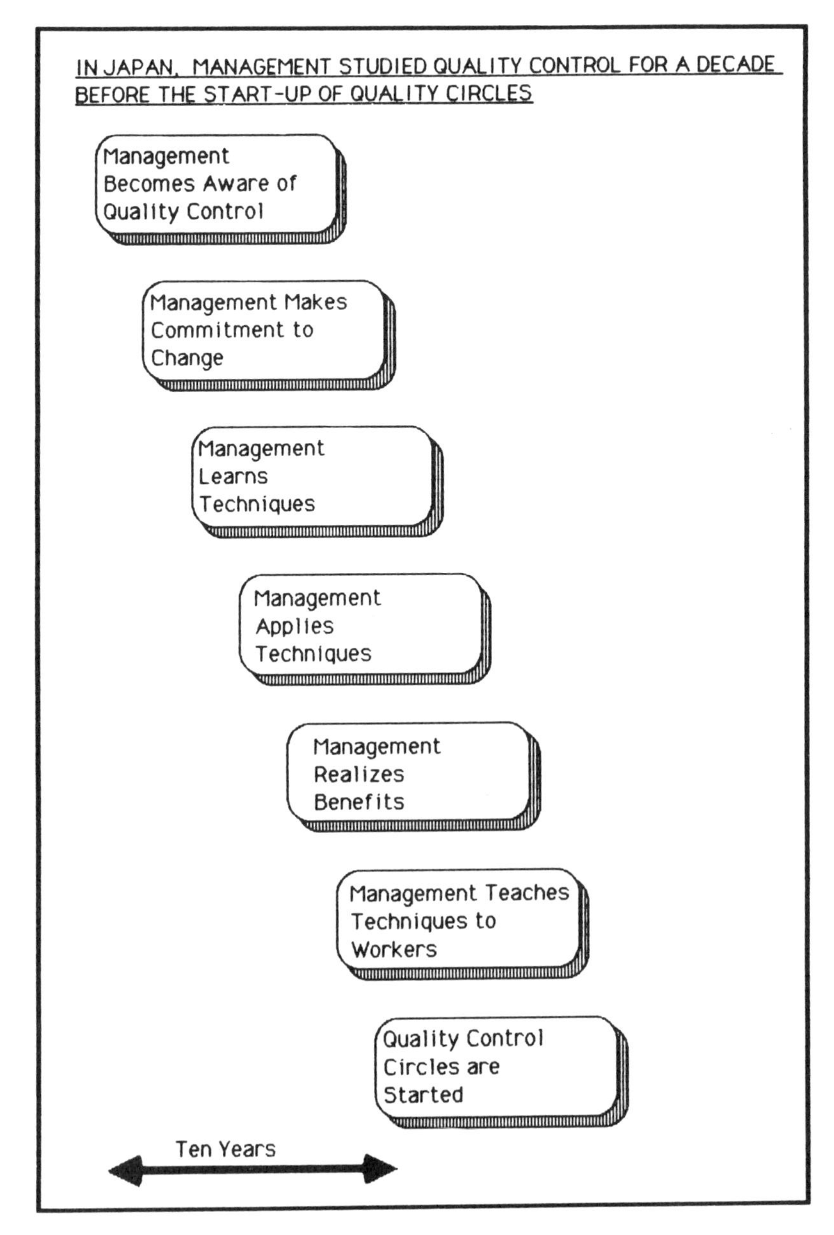

Figure 9.5 What happened in Japan before quality circles.

He goes on to state, "A fourth view that is quite widespread among western observers is that Japanese unions are highly supportive of circle activities. The facts are that in most cases the unions are, at best, passive observers" [8].

While there are relatively few unions in banks, some experience of unions vis-a-vis QC circles have been reported in unionized service industries. For instance, the union in the Depository Trust Corporation chose to make the few QC circles of DTC a topic of their negotiation. They wanted the circles discontinued. The apparent feeling was that they (the union) felt that circles were exploiting the workers. Apparently, DTC did not get a great deal of involvement of the union when starting their circle activity.

Professor Cole goes on to note that:

> A final misconception which I would like to address is that the Japanese have pretty well solved the problems of worker motivation and that the circles are now self sustaining activities. The situation is rather that the Japanese are constantly concerned with mannerism or *mannerika* as they call it. By this they mean that there is a constant problem of circles reverting to ritualistic activity and they are continually struggling to combat this. . . .
>
> This last point brings us to a consideration of the fact that QC circles do not work uniformly well in Japanese companies. There is a tremendous variation in the success of these activities. Even in the supposedly best companies, management will commonly acknowledge that the circles are working well in less than half the cases. [9]

Quality control circles have a place in Japanese industry. Their place is in an integrated quality control program driven by top management. It is important that this be kept in mind. To show the place of quality control in the scheme of things, it is well to examine the Japanese concept of "Total Quality Control."

Used As Part of Total Quality Concept

The Japanese concept of total quality control is more extensive than the concept applied in many American firms. It is based on the Deming principles, which are just now catching on in the United States. Dr. Sandholm quotes Dr. Ishikawa in an outline of the meaning of total quality control that is worth noting:

Ishikawa gives six features of quality work in Japan [10]:

- Company-wide quality control
- Quality control audit
- Industrial education and training
- QC circle activities
- Application of statistical methods
- Nation-wide quality control promotion activities

Company-wide quality control—This refers to Deming's Point 14. above. Everyone in the company is involved in a positive manner in the process of quality. Top management sets policy for quality as it is set for finance in the United States. Deming, in 1950, taught the Japanese the Deming cycle: 1. design the product, 2. make the product according to the design, 3. sell the product, 4. do research to get customer's and noncustomer's comments on the product for improvement, 5. go to step 1. [11]. (See Figure 9.6.)

Company-wide quality control includes the principle of constancy of purpose. The company intends to stay in business. A bank in order to stay in business must constantly invent new ways of obtaining funds such as CD's, BA's and more recently mortgage instruments. They must also constantly find new ways to service their customers which means principally new methods for investments and loans.

Quality control audit—This has a different meaning in Japan. In the United States a quality audit is most often a review of the product and procedures pertaining to the quality of what is being made. This audit is usually conducted by a team of staff personnel who can do no more than observe and report. In Japan the audit is performed by executive management of the firm. Often the president heads the audit team.

The result of having executive management audit the system is enormous. They get first hand knowledge of problems and are in a position to remove the barriers that prevent workers from achieving quality. They are in fact implementing a number of Deming's 14

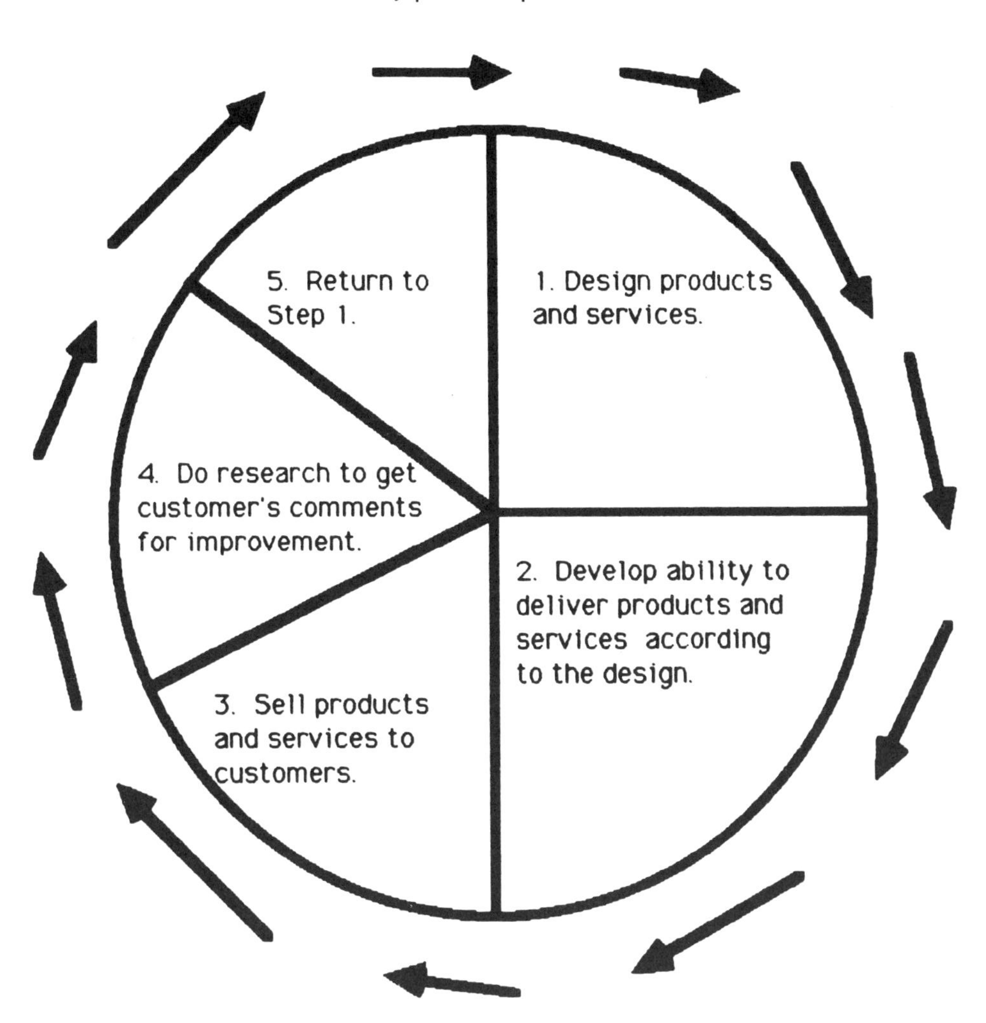

Figure 9.6 The Deming cycle.

Points. At the same time, they are signalling to the firm their commitment to the quality policy and plans.

Industrial education and training—A key element of Japanese quality programs is the continuous education and training of the whole work force (Deming's Point 13.). A workforce that knows the concepts of quality control can apply it on the spot provided the other 14 Points are in place. At least one large regional bank is looking at starting such a program in 1985.

QC circle activities—Dr. Sandholm gives a good insight into why Japanese QC circles contribute only 10% to the improvement of quality:

> Why is such a small fraction of the improvements attributable to the quality circles? In Japan, circles tackle interdepartmental [*sic*] problems almost exclusively. These usually represent the "trivial many" problems. The "vital few" problems are difficult for the workers to tackle in quality circles because these problems are caused by absence of management policies, bad quality coordination, insufficient training, design weaknesses, incapable vendors, etc. [12]

From the context of the paragraph it seems that Dr. Sandholm means *intra*departmental problems rather than *inter*departmental problems. The intradepartmental problems are the trivial many, the small problems that lie within the scope of solution for the quality circle. A quality circle to be truly effective cannot consider problems over which it has no control. (See Figure 9.7.)

Application of statistical methods—Until a problem can be defined no solution is possible. To define a problem it is necessary to first eliminate what are called the non-random effects. Non-random effects are those things operating on a system which cause wide, out-of-control variations to occur. For instance, ill-defined jobs cause operators to do the job one way or another without plan or reason. The result is a series of mistakes. Until the job is defined, no one can predict the error rate which is sometimes good, sometimes terrible. Big penalty payments result.

Deming taught the Japanese to remove the non-random errors, then measure the random errors and make improvement. Juran taught an organized method for improvement, the managerial breakthrough technique. During the time that improvement is being effected, the operation remains under close control. It is delivering an output that is constant and predictably so.

THE AMERICAN EXPERIENCE WITH QUALITY CIRCLES

At first, American Quality Circles were able to address relatively simple problems that had big benefits.

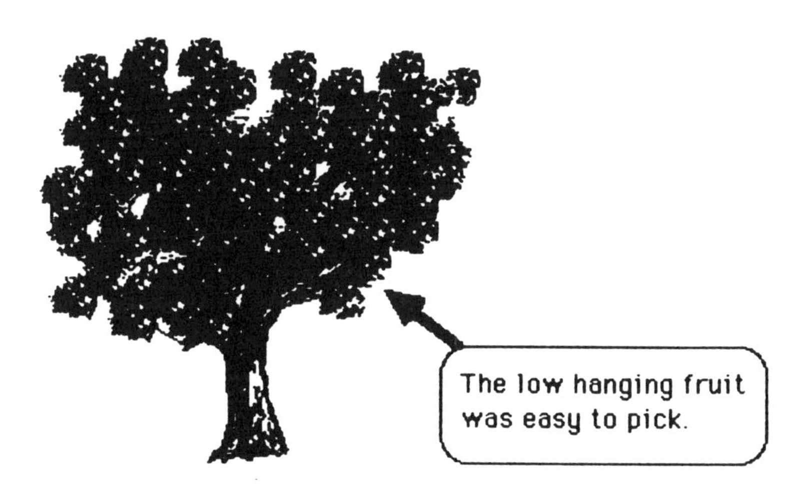

Later, the Circles faced the "trivial many" problems. The "vital few" were beyond the scope of their control.

Figure 9.7 A view of the American experience with quality circles.

The tool used to attain and maintain control is the control chart. This tool allows management to know what is going on in the process. It provides the operational definition of the process. It is the basis from which improvement is possible.

Nation-wide quality promotion activity—Japan has recognized the need for a national effort for quality control. The initial impetus came from the Union of Japanese Scientists and Engineers also known as JUSE from their telegraphic acronym. They invited Deming in 1950. They recorded his lectures and published them. When they wanted to give the royalties to Dr. Deming, he generously gave them to JUSE. JUSE used the royalties to found the Deming Prize for quality.

The Deming Prize is awarded in November. JUSE was able to get national attention and to have the prize awarded by the Emperor personally. The result was a tremendous impetus on the part of corporate officials to win the prize. It carries with it national prestige.

In October of 1984, the American Society for Quality Control succeeded in having a bill passed in Congress declaring October a quality month. It is a good start. But the American Deming medal is unknown except to a handful of people.

The Japanese idea of presenting national prestige to those leading in statistical quality control has greatly contributed to its advances in Japan. It remains to be seen whether the efforts in the United States will bear the same result.

The American Experience

Great care must be exercised in comparing Japanese and American situations; the conditions are different. The concept of corporate ownership is substantially different. As Dr. Myron Tribus observed after a research trip to Japan:

> In the Western world the company's physical assets are considered to be the "company" and to belong to the stockholders. In this view the company is an economic unit. The managers and workers are "hired hands" who are put on the payroll to enhance the value of the stockholders' investment.

> Therefore, whenever a project is considered a calculation of the return on investment (ROI) is made. Unless the ROI is high enough to compete with alternative investments that could be made by stockholders, the company is reluctant to make the investment.
>
> In Japan, in contrast, the stockholders are considered to be silent "partners" of the managers and workers. The company is considered to be the managers and workers, not the equipment. The value added by the company is computed by the simple equation:
>
> $$\text{Value added} = \begin{matrix}\text{Value of all goods}\\ \text{and services sold}\end{matrix} - \begin{matrix}\text{Cost of all goods}\\ \text{and services purchased}\end{matrix}$$
>
> This value added is available to be shared by the partners. One of the speakers at the [International Productivity] Symposium [Tokyo, May 12-14, 1983] suggested the division 1/3 to the shareholders, 1/3 to be reinvested in the company and 1/3 for the managers and workers. [13]

In Japanese firms, the pressures driving management are different. In the United States it is shareholder equity; in Japan it is market share. (See Figure 9.8.)

Role of Top Management

While stockholder pressure is real in the United States, that is not the real reason why so few companies are following the Deming Principles. It should be noted that a number of very large U. S. firms are following the Deming method used by the Japanese with equally good results.

The real problem is management commitment. As Dr. Tribus observed:

> It is often said that the Japanese culture makes a difference. Certainly there are differences, but some of the differences are in the "corporate culture," which is determined to a large extent by the leadership inside the company, rather than the culture outside the company. Company cultures do not change overnight. One CEO put it this way: "The President should not be faint hearted." [13]

THE UNITED STATES AND JAPAN HAVE DIFFERENT VIEWS OF PROFITS

In Japan:
- stockholders are silent partners to management and workers,
- the company "belongs" to managers and workers,
- profits are considered value added.

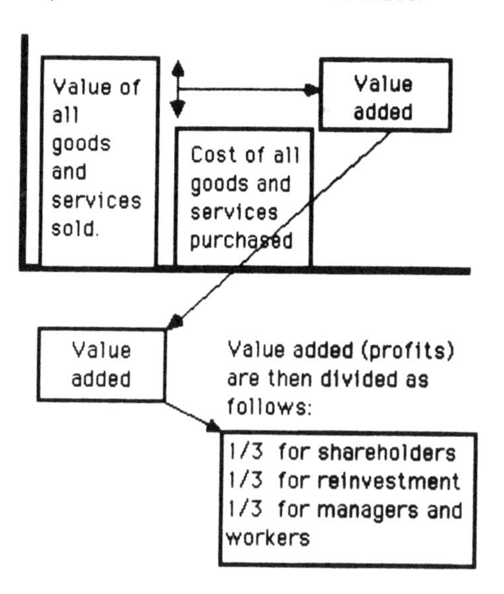

Management is driven by market share.

In the United States:
- the company "belongs" to the stockholders,
- "We milk the cow too often !!!"
- profits are considered return on investment.

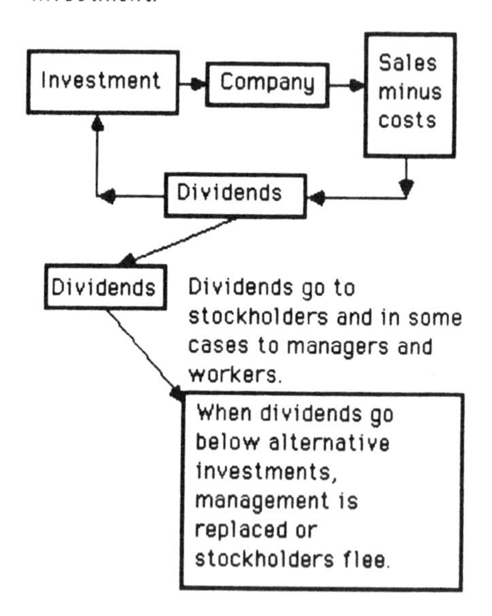

Management is driven by quarterly profit and loss statements.

Figure 9.8 The view of stockholders in Japan versus America.

As stated several times, in an attempt to emulate the success of Japanese management, American managers turned to QC circles, the most spectacular and visible of the Japanese methods. Those that studied the QC circle in Japan were told of the many successes that the movement had. One can suspect that the failures were quietly avoided. As Dr. Goodfellow stated, "While success has a thousand fathers, failure is an orphan" [14].

Quality circles are easy on management. Simply install them and problems of quality are supposed to go away. Executives can think that they can pay attention to the other demands of business once the quality circles are installed. Unfortunately, that is not what happens. In many cases the situation grew worse. Why? Because "Quality Circles Won't Work Without Quality Control" [15].

> Quality circles which involve tapping the ideas of white collar and non executive personnel to improve workflow and productivity, have become the talk of the banking industry. At least a half-dozen major banks have begun using them. Several more are expected to follow suit next year [1982].
>
> Unfortunately, many of these circles will probably fail to fulfill their promise, for one simple reason. They are set up before their banks achieve true quality control. Quality control means both the ability to turn out product of consistent, known quality and the ability to measure that quality.
>
> When such control is not in place, quality circles can run into or create numerous problems. First, they may not obtain the support of top management, and no program for improvement can go far without such support. Second without quality control, quality circles have no way of assessing the importance of problems or seeming problems in the workflow or productivity. As a result they may treat as major problems what are merely occasional fluctuations in quality. Third, they have no way of measuring the impact of their own ideas for improvement.

Quality circles are generally composed of a group of 10 people who work together. Their supervisor is often a member of the circle. Many times a facilitator, someone from outside the organization also attends to help keep the circle moving. Assuming an average salary and overhead of $10.00 per hour, 10 people meeting for 50 weeks cost the company about $5,000 per year. One bank reported having 2,500 people involved with a savings over a two year period of $200,000 [16].

Assuming the numbers above, the return on an investment in labor of about $2,500,000 over two years yielded savings of $200,000. Less than 10% of the investment. And this for a bank that has an excellent quality control system in their operations area. If they had extended their excellent quality control to the lending area as well, they might not have experienced the disaster that required reorganization of the bank under FDIC.

Whenever one discusses the negative return on investment that quality circles seem to experience in this country, the argument is made that there are many other benefits in human relations which are achieved and which are not measurable. While there is much validity in this argument, it implies that the main use of quality circles is not improved quality control as top management thought but making up of another management deficiency, lack of good employee relations. The question is whether the investment in time and energy is worthwhile or if the same impact on human relations could be achieved better in less expensive ways.

Can workers solve problems of quality? Yes, if the problem is controllable by them, if they are adequately trained for the task, if management is following Deming's 14 Points and if they are properly guided. Four major if's. Unfortunately, the conditions are rarely met in practice.

Many quality circles are "installed." Sometimes with the expert help of consultants. In most cases, the use of quality circles is a substitute for a quality control program. Many of these circles arrive at the same recommendation: "We need more training." This recommendation is probably justified. In most cases errors are occurring because the workers job is not completely defined or lines of responsibility are fuzzy.

Role of Middle Management

In many cases the upper management is "sold" on the concepts of QC circles. Workers like it because, at first, it promises them a voice in their work place. They are now free to make suggestions directly to upper management. It would seem that all is well, but it is not so.

Middle managers see QC circles as practiced in many companies in the United States as a distinct threat to their authority. If workers can make suggestions directly to upper management on how to improve the operation, then what is their, middle management's, role. At best they will look silly for not having had the idea in the first place, at worst they will be blamed for having missed an opportunity.

Under the circumstances, it is not surprising that middle management feels resentful towards QC circles. They share this feeling with union leaders. It is not unknown that QC circles have failed due to middle management influences.

Role of Supervisor

Dr. Goodfellow studied 29 American companies that had quality circle activity. In only 8 of these did he find circles that could be described as successful. By comparing the 21 unsuccessful circle endeavors with the 8 successes he was able to arrive at some conclusions of the basis for success [14].

His key finding was that the training of the foreman as a circle leader was the most significant condition for success. Among such training characteristics as fitting the training to the situation, the major point was "... the best kind of training to achieve those ends [employee motivation] is training the foreman *how to listen*. That was the secret of the foremen's success in the eight successful plants" [17].

There were two aspects to success, one was the foreman who listened with understanding and the other was the ability to get things done for his men by "higher management":

> If an employee can turn to a foreman who is not "critical," who is not "hard to talk to," and who can help with "problems" — if, in short, the foreman is a good listener and *tries to do something* in response to a production problem, he will receive excellent cooperation from his work force in their QCC goals. [18]

Why do foremen or supervisors not listen? Part of the reason is that they fear that their authority will be undermined if they accept a suggestion from the work force. The Taylor concept, nourished over the years has taught supervisors that they, not the workers, were the "smart ones." If the supervisor had to turn to help from his employee, he was not fit to do his job. Suddenly, management is reversing this and asking the supervisor to cooperate with his people in a QC circle activity. Supervisors need to be trained to do this task.

The question might legitimately be raised that if it is a question of training supervisors to listen and act on suggestions, why not simply do this and save the expense of quality circles. Apparently the impact is the same. In fact, the application of Deming's 14 Points achieves this effect. While there appears no record of a bank having used this method, other Service Industries have successfully done so [19].

Managerial Breakthrough Technique

Apparently the whole concept of QC circles is based on a Japanese version of Dr. Juran's Managerial Breakthrough Technique (MBT). It is interesting to watch an attempt to re-import a method and fit it to American use when the original method is very much available and yet little used. Like QC circles, it too can be misused when it is substituted for a proper program of quality management. However, used in the proper context, MBT is a powerful management tool.

Original Juran Concept

Published in 1964 after more than a decade of research, Dr. Juran's book, *Managerial Breakthrough* is still an important text today. It deals with improvement. Dr. Juran found that improvement was achieved through a sequence of events that are constant regardless of the industry in which they are applied.

Dr. Juran points out that most quality activity today concentrates on getting a system under control, that is, having the system produce a continuous stream of product that is of a known, consistent quality. Thus a bank might have a stable, consistent MICR reject rate of 4%.

But a reject rate of 4% is not a good reject rate. True all special causes for this high rate have been eliminated. What are the steps to reduce this reject rate? Dr. Juran gives the answer [20].

The steps are:

Breakthrough in attitude
Use of Pareto principle [to select areas for research]
Organizing for breakthrough in knowledge
Creation of a steering arm
Creation of a diagnostic arm
Diagnosis
Breakthrough in cultural patterns
Transition to new level

Management Action

Breakthrough in attitude—The first step in the sequence is the realization and desire to improve. When senior management finds that a consistent 4% reject rate is too high, they want to do something to reduce this rate and form a group assigned to this task, a breakthrough in attitude has been achieved. Just saying something should be done about the reject rate is not enough. Management must indicate by action that they are interested enough to back the venture.

The main way for management to show this commitment is in the formation of a task force to do the job.

The Pareto analysis—There are probably many sources to the problem which combined give the effect. Perhaps when reviewed it is found that there are many different kinds of checks contributing to the rejects. Payroll checks, accounts payable checks, other corporate checks, internal tickets, personal checks among others all have a different contribution to the error rate. By ranking the contribution from high to low and drawing a bar graph with the highest contributor on the left, a Pareto chart is established.

The Pareto principle is that a few cases contribute most to the problem under study. Some say that 80% of the problem comes from 20% of the cases. That 20% Juran calls the "vital few." It is obvious that more benefit can be gained by working on the vital few than by working on the remainder which he calls the "trivial many."

The Pareto analysis, *viz.* ranking the contributors to the problem and selecting the major contributors, is a way of establishing priorities to be worked on.

Method and Results

Organization for breakthrough in knowledge—In order to accomplish anything, the project must be sanctioned and a group appointed to carry out the task. The organization proposed by the technique uses two groups, 1. a steering committee and 2. a diagnostic arm. Depending on the complexity of the problem, these groups may be large interdepartmental units or a small group of a single department. They are generally large since the vital few are major problem areas to the organization.

The steering arm—This is the controlling group. It generally consists of managers who have responsibility, either directly or indirectly, related to the problem to be examined. This group sets the policy of the research through developing potential causes which need to be explored. Based on the result of the research they commission, they also develop potential solutions for implementation. The group should have sufficient responsibility to be able to implement their solution. This argues for an executive body rather than a low level group.

The diagnostic arm—The actual research is carried out by the diagnostic arm of the group. Technically trained people will be needed to make the requisite measurements and observations that can furnish the facts to the steering committee.

The diagnostic arm carries out the studies indicated and comes up with what Dr. Juran calls the "breakthrough in knowledge — diagnosis" [21].

Breakthrough in cultural pattern—It is the breakthrough in knowledge that permits the steering arm to arrive at the changes needed to effect solution. However, Dr. Juran points out that the mere knowledge of what needs to be done is often not enough. It is important for the steering committee to be cognizant of the organization culture. In order to successfully accomplish change, the organizational culture must be taken into account. Provision must be made for this, or the project has a good chance of failure.

Breakthrough in performance—Given the foregoing steps, it is now possible to implement the change needed to transit to improved conditions. For instance, if it is found in the example of reject rates that the checks from certain vendors are a major contributor to the reject rate, that these checks are purchased by the bank for writing dividend payments and that the check order is placed four days before the checks are needed, a number of changes must be made to improve the process. The reason the vendor was bad was because of the short lead time. The short lead time was used because, "we always did it that way." It was an unnecessary thing to do. The steering committee convinced the Corporate Trust Group that a lead time of five weeks was possible. This allowed better printing and a drastic reduction in reject rate.

Summary and Conclusions

As can be seen from the brief description given above, the solution to problems of the system can be approached in a structured fashion. They can only be approached when the system is in control so that the diagnostic data gives trustworthy results. They can only be ordered and implemented by management. Quality circles are impotent to operate on interdepartmental problems which is where the "vital few" are to be found. They are, however, an excellent resource for work on intra departmental problems. The "trivial many" problem causes are often found contained within departments and are a proper area for quality circle concern.

References

1. Joseph M. Juran, "Section 18 Motivation," in J. M. Juran, F. M. Gryna, Jr., and R. S. Bingham, Jr. (Editors), *Quality Control Handbook*, Third Edition (New York: McGraw-Hill Book Company, 1979), p. 24.
2. According to W. E. Deming this was originally stated by J. M. Juran and is often cited. W. J. Latzko measured this amount and found that management had more control than 85% (see W. J. Latzko, "Quality will determine winners in competitive banking market," *Bank System & Equipment*, March 1984, p. 204; J. M. Juran and F. M. Gryna, Jr., *Quality Planning and Analysis*, Second Edition (New York: McGraw-Hill Book Company, 1980) p. 107, and Jeremy Main, "The curmudgeon who talks tough on quality," *Fortune* (June 25, 1984) p. 119.
3. W. Edwards Deming, *Quality, Productivity, and Competitive Position* (Cambridge: Massachusetts Institute of Technology, Center for Advanced Engineering Study, 1982), pp. 17-50.
4. Donald L. Dewar, *The Quality Circle Guide to Participation Management* (Englewood Cliffs: Prentice-Hall, 1980), p. 11.
5. Lennart Sandholm, "Japanese Quality Circles—A remedy for the West's Quality Problems?" *Quality Progress*, Volume XVI (February 1983), p. 21.
6. Robert E. Cole, "Common Misconceptions of Japanese QC

Circles," in *Thirty-Fifth Annual Quality Congress Transactions* (San Francisco: American Society for Quality Control, 1981), p. 189

7. From Kaoru Ishikawa, "Japanese Total Quality Control," part of a panel discussion presented at the American Society for Quality Control's 36th Annual Quality Congress, Detroit Michigan, May 4, 1982.
8. Robert E. Cole, "Common Misconceptions of Japanese QC Circles," in *Thirty-Fifth Annual Quality Congress Transactions* (San Francisco: American Society for Quality Control, 1981), p. 188.
9. Robert E. Cole, "Common Misconceptions of Japanese QC Circles," in *Thirty-Fifth Annual Quality Congress Transactions* (San Francisco: American Society for Quality Control, 1981), p. 189.
10. K. Ishikawa as quoted in Lennart Sandholm, "Japanese Quality Circles—A Remedy for the West's Quality Problems?" *Quality Progress*, Volume XVI (February 1983), p. 20.
11. W. Edwards Deming, *Elementary Principles of The Statistical Control of Quality*, Second Printing June 1952 (Tokyo: Nippon Kagaku Gijutsu Remmei, 1951), p. 9.
12. Lennart Sandholm, "Japanese Quality Circles—A Remedy for the West's Quality Problems?" *Quality Progress*, Volume XVI (February 1983), p. 21.
13. Myron Tribus, "Reducing Deming's 14 Points to Practice," an unpublished report enclosed in a letter from Dr. Myron Tribus, Director CAES-MIT, June 13, 1983, pp. 12-13.
14. Matthew Goodfellow, "Quality Control Circle Programs—What Works and What Doesn't," *Quality Progress*, Volume 19, Number 8 (August 1981), p. 31.
15. William J. Latzko, "Quality Circles Won't Work Without Quality Control," *The Magazine of Bank Administration*, Volume 55, Number 12, December 1981, p. 23.
16. Lawrence A. Eldridge and Charles A. Aubrey II, "Stressing Quality—The Path to Productivity," *The Magazine of Bank Administration*, Volume 57, Number 6, June 1983, p. 22.
17. Matthew Goodfellow, "Quality Control Circle Programs—What Works and What Doesn't," *Quality Progress*, Volume 19, Number 8 (August 1981), pp. 31-32.

18. Matthew Goodfellow, "Quality Control Circle Programs—What Works and What Doesn't," *Quality Progress*, Volume 19, Number 8 (August 1981), p. 32.
19. From William J. Latzko, "Roadmap for Change," a speech presented at a meeting of the Stamford, Ct. Section of the American Society for Quality Control, January 9, 1985.
20. Joseph M. Juran, *Managerial Breakthrough* (New York: McGraw-Hill Book Company, 1964), p. 14.
21. Joseph M. Juran, *Managerial Breakthrough* (New York: McGraw-Hill Book Company, 1964), see Chapter 8.

10
Summary

This book reviewed the principles of quality control for bank operations. In passing there were comments referencing the many other areas where quality control has applicability.

Banks have thought that their problems are in their operations. "If only the operators would not make so many mistakes," is a common statement heard in banks. The facts are really different. The most costly mistakes are not made in operations, they occur in lending and managing the bank. Blaming the operations just frustrates operating personnel and hides the true issue. Blaming the operations guarantees the perpetuation of the problems compounded by higher turnover and increased problems.

Clever bank management will use this text to beat their competition. The way is clear. When the real concepts of quality management are embraced by the executives of the bank, they will beat the competition with better customer oriented products delivered with consistently good quality. The banks that follow this route will be the survivors.

Today there are about 14,000 commercial banks. There are many other banking concerns or financial intermediaries as they are sometimes called. Savings banks, credit unions, insurance companies, retail corporations, oil companies, brokerage firms among others are all competing for banking business. The 14,000 commercial banks will shrink to perhaps 200 in the next decade. Successful banks will absorb the losers. Who will be the winners? It will be those banks that are managed on sound principles such as Dr. Deming's 14 points.

In spite of its usefulness in lending and managerial areas, this book concentrated on the aspect of operations. The reason is simply that the principles are more easily demonstrated in the operations areas. In addition there are some immediate benefits possible in such mechanized areas as checks processing. This and similar areas are really small manufacturing processes. The methods of quality control used in factories can be used in these operations with little or no change.

One word of caution is in order. The mere use of measurements does not give results. Management must be an active partner in the process to fix the problems disclosed by the measurements. Executives have the key to success in their hands. Will they use it?

A bank that is interested in starting a program of managing quality has to select the place to begin. Some pointers on that topic are covered in Chapter 8. The selection of a starting point is an art. A bank considering this move is well advised to get qualified help. In the long run this will save them time and money. A poor start can permanently eliminate any consideration of the use of the tool of quality management.

The bank should also consider the structure under which the system can work. In a number of cases, the quality management is operated by a senior bank executive. The level varies. It is interesting to observe that those banks which are most successful in establishing an effective system appointed a senior vice president to be in charge reporting directly to the president or chairman.

The use of a high level executive with direct access to the top gives an important signal to the whole organization. No one in the bank can question management's intent; the signal is clear.

The most effective organization is to have a very small central staff for the coordination of all efforts. The actual work of implementing the quality process should be in each area. Depending on the size of the bank, technical expertise can be internal or consultants can be used. The important thing is that everyone in the organization be trained to apply the principles of quality management. This means that managers from the supervisory level up have a full understanding of the their job and how to use quality control.

A bank whose management team is trained can cope with the day to day problems and achieve better quality and productivity than if the team is not trained.

The training is different for varying levels of management. The mix of philosophy and technical skills vary. At each level the managers must have the full knowledge of how to make the system work, how to tell whether it is working or not and how to fix it if there is a problem.

This type of training has not found its way into the school systems nor is it taught in business schools. Those banks that have established such training programs have done so by calling on skilled help from within (where it exists) or hiring outside qualified personnel. Again there is a note of caution: not every one who claims to be qualified is qualified. Practical banking or related experience is of value that cannot be underestimated.

Once the organization and starting point have been established there are a number of techniques available. These are described in the Chapters 2 through 5. Essentially, use is made of statistical methods found useful in wide application of problem solving. These methods are tailored to the non-manufacturing way of functioning in a bank. By adopting the main theory of quality control to the special needs when studying people-driven operations, a useful tool is formed.

The principle tool from the technical side is the control chart. Since it was invented half a century ago, the use of the control chart has enabled skilled managers to improve quality and productivity. Facts are the tools needed to manage. In the absence of facts, the glibbest person will get the largest budget. The control chart answers the question of the source of the problem and the extent.

There are some economics to the quality process. These, however, can be deceiving. In Chapter 6 the concept of economics is covered. It must be noted that the only costs reviewed are those that are measurable. There are many costs that cannot be measured directly but are due to the quality of the work. For instance, customer satisfaction is not entirely measurable by share of market. The relationship is not direct.

There is a major cost impact of quality which cannot be measured.

If the work coming to a clerical department is clean, that is to say contains no missing or ambiguous data, it is easier to use. This gives a boost to worker morale and productivity.

For a time banks relied heavily on quality circles to solve their operational problems. This worked in some banks such as Continental Illinois and Irving Trust because there was a strong foundation of quality management underlying the use of quality circles.It was uniformly successful. Unfortunately, the failures were due to the misapplication of the method rather than the method itself. Chapter 9 addresses the topic in some detail.

Chapter 9 also contains some features of a quality philosophy as well as a very brief description of the approach for quality improvement. Substantial benefits can be realized from both of these topics.

While banks have not been the leaders in the use of modern methods of quality management, there has been a great deal of interest shown by them. A number of banks pioneered the process. These were large money center banks who were able to take advantage of developing methods that materially improved their operations and gave them an edge over the competition. Now that the methods are developed, all banks have the benefit of their use.

The banking system, especially the commercial banks, are locked in a competitive struggle. The use of quality management will determine the winners.

Appendix 1

Copy of Bank Quality Control Survey

The following is a copy of the report prepared by Mr. Eugene Kirby of the status of quality control in banks.

COPY

American Society for Quality Control
Administrative Applications Division
Banking Committee

Subject: Result of Recent Survey of Banks

In January of 1975, survey forms were sent to approximately 50 banks requesting information on Quality Control. These 50 banks were selected from a list of the 100 largest banks in the country.

The results of the survey are attached.

Thank you for your participation.

/s/ Gene Kirby
Eugene Kirby
Chairman, Banking Committee

TABLE A1.1 Summary of Findings

Size of Bank	Number Responded	Formal QC	MICR QC	Product QC	Project QC	Data Security	Audit Approval	Member ASQC
No Formal QC Function	4	No	—	—	—	—	—	—
$1B–$3B	2	Yes	Yes	Yes	Yes-1 No-1	Yes-1 No-1	Yes	Yes
$3B–$5B	2	Yes	Yes	Yes	Yes-1 No-1	No	Yes-1 No-1	Yes-1 No-1
$5B–$10B	3	Yes	Yes-2 No-1	Yes	Yes-1 No-2	Yes-1 No-2	Yes-1 No-2	Yes-2 No-1
Over $10B	4	Yes	Yes-2 No-2	Yes	Yes-2 No-2	No	Yes-3 No-1	Yes-3 No-1
Percent	30%	72%	53%	72%	33%	13%	46%	53%

Additional Findings for Table A1.1

In addition to those banks [in the table] indicated as responding, four banks responded "No" to the question of a formal QC function. No additional information was offered (15 responses of 50 inquiries).

All banks responding have some form of production control. New York banks are particularly involved with clerical output. Eight [of these] banks reported formal MICR QC programs — the largest area of commonality among the banks.

Two banks report Data security *as a part of QC*. This does not necessarily mean that the other banks have no Data Security effort.

Eight banks reported that Auditing has approval authority on applications prior to turnover to Operations.

Appendix 2

Derivation of Equations to Determine Inspector Efficiency

To estimate the probability of not finding a defect after n levels of inspection, it is necessary to determine the probability of failing to find a defective item after n inspections and the probability that an item is defective in the first place. This is given by the conditional probability statement:

$$P\ (D\,|\,n) = P\ (F\,|\,n) \times P_0(D) \tag{1}$$

To arrive at this probability statement, it is required to derive the two components.

Probability of Failing to Find a Defective Item

It is acknowledged that checkers are not perfect. The probability of failing to find a defective item is merely a quantification of the degree of imperfection that one can expect from one or more checkers examining the same item. The following definitions are used:

I = Errors detected by the checker (Internal Failure);

E = Errors not detected by the checker (External Failure);

F = The ratio of failing to find errors which are present;

$$F = E/(I+E) \tag{2}$$

n = The number of checkers examining the same item.

The values of E and I can be obtained from error data. For instance, items returned by the customer for correction are the base for estimating the value of E. The errors found internally from QUIP data are a good estimate of I.

It is easy enough to see what happens when a single checker is involved. However, when two or more checks are performed in series the situation changes slightly. To see the effect, consider two checkers examining the same set of documents. If there are 100 defects among the documents, and the checkers are able to catch 80% of the defects, the first checker will remove 80 defects leaving 20 in the lot. The second checker will presumably find 80% of the 20 defectives, leaving 4 items to escape.

Each checker in the example failed to find 20% of the defects ($F = .20$). As a result, the four items out of the 100 escaped:

$$100 \times .20 \times .20 = 4$$

or in general

$$(I + E)(F_1)(F_2) = E \tag{3}$$

This equation (3) can be extended to any number of checkers:

$$(I + E)(F_1)(F_2)\ldots(F_n) = E \tag{4}$$

If the failure rate of each checker is the same as every other checker, then $F_1 = F_2 = \ldots = F_n = F$. In that case equation (4) simplifies to:

$$(I + E)F^n = E \tag{5}$$

By a simple rearrangement, the failure rate for n checkers, or, the probability of failing to find a defective item after n levels of checking is

$$P(F \mid n) = F^n = E/(I + E) \tag{6}$$

If there are two levels of checking and over a period of time it was observed that although some 4,700 defective items were caught and corrected, 300 defective items were returned by the customer. This gives a total of 5,000 defective items produced.

$$(I+E) = (4{,}700 + 300) = 5{,}000$$

The probability of failure given two levels of inspection in this case is

$$P(F|n) = P(F|2) = F^2 = 300/5{,}000 = 0.06$$

The failure rate for a single checker is the square root of 0.06 in this illustration. This value of F is 0.245, or 24.5% of the defective items are generally missed by each checker.

The Probability That an Item Is Defective

The probability that a particular item is defective can be estimated by the relation:

$$P_0(D) = (I + E)/V. \tag{7}$$

In this equation V is the total volume processed. If in the example above the total volume processed in the time period is 400,000 then

$$P_0(D) = (300 + 4700)/400{,}000 = 0.0125 \text{ or } 1.25\%.$$

This is the quality of the work prior to inspection. The inspection process is designed to improve this level of quality.

Efficiency of n Levels of Inspection

The efficiency of varying levels of inspection can be determined by use of equation 1. In the example described above, the following result is obtained:

$$P(D|n) = P(F|n)P_0(D) = (.245)^n \times 0.0125$$

For three levels of checking this value turns out to be 0.000184 or about one defective in 5,500 items defective are missed on checking.

In working out the statistics for checker efficiency, a number of significant assumptions were made. If these assumptions do not hold, or do not hold as well as it is liked, the probabilities are adversely effected so that the results will be less efficient than the pure theory allows.

The first assumption is that of inspector independence. This means that it is assumed that the presence of an inspector who will check the same work after the current inspection has no influence in the current inspector. Likewise, the fact that an item has already been checked is assumed to have no influence on the next inspector. This theory of independence is not as strong as is desired in practice. It is particularly influential if the inspector is working under stress, such as trying to meet deadlines or having too large a workload. In such a case there may be a tendency to relax the inspection process on the theory that the "other" inspector will catch what the present inspector misses.

The second assumption is that the number of defective items in a lot presented for inspection does not affect the inspection efficiency. Experience shows that this assumption becomes weaker as the number of defective items becomes either a very large percentage of the lot or a very small percentage. In either case, the inspection efficiency seems to decrease.

The third assumption is that the pressure of the work does not impact inspector efficiency. The mere pressure of a lot of waiting items can have an influence on the quality of the inspection.

The fourth assumption is that the inspector efficiency is constant throughout the day. Factors of fatigue are neglected in the theory.

The fifth assumption is that the inspectors have at their disposal all the tools they need to perform the inspection and do not have to rely on production tools. Thus the inspectors must have the latest updated procedures and standing instructions in order to do a proper job.

Bibliography

Books

Adam, E. E. Jr., Hershauer, J. C., Ruch, W. A., *Measuring the Quality Dimension of Service Productivity*, National Science Foundation Grant No. APR 76-07140, January 1978.

American Insurance Company. *Quality Improvement Techniques*, New York: American Management Association, 1962.

American Society for Quality Control. *Quality Costs—What & How*, 2nd edition. Milwaukee, Wisconsin, 1971.

American Society for Quality Control. *Quality Motivation Workbook*, Milwaukee Wisconsin, 1967.

Apps, E. A. *Ink Technology for Printers and Students*, Vol III. Cleveland, Ohio: Chemical Publishers, 1964.

Bergstrom, James. *Teller Differences Rate. A Study of Factors Analysis*, 4th edition, Philadelphia, PA: ASTM Special Technical Publication 15D, 1976.

Bergstrom, James. *Teller Differences Rate. A Study of Factors Affecting Teller Performance*, Publication 700, Park Ridge, Illinois: The Bank Administration Institute, 1976.

Bowker, Albert H., and Gerald J. Lieberman. *Engineering Statistics*, Englewood Cliffs, New Jersey: Prentice-Hall Inc., 1959.

Brech, E. F. L. *Management, Its Nature and Significance*, New York: Pittman Publishing Company, 1967.

Caplen, Rowland. *A Practical Approach to Quality Control*, Princeton, New Jersey: Brandon/Systems Press, 1970.

Eastman Kodak Company. *Control Procedure in Microfilm Processing*, Rochester, New York, 1974.

Edwards, Chilperic. *The Hamurabi Code*, reissued 1971. Port Washington, New York: Kennikat Press 1904.

Committee E-11 on Statistical Methods, American Society for Testing and Materials. *ASTM Manual on Presentation of Data and Control Chart Analysis*, 4th revision. Philadelphia, Pennsylvania: ASTM Special Technical Publication 15D, 1976.

Control Chart Method of Controlling Quality During Production — ANSI Standard Z 1.3-1975, New York: American National Standards Institute, Inc., 1975.

Cope, Frank E. *Quality Control at Gulf Life Insurance Company*, Gulf Life Insurance, 1971.

Corns, M. C. *The Practical Operation and Management of a Bank*, 2nd edition. Boston, Massachusetts: Bankers Publishing Company, 1968.

Cowden, Dudley J. *Statistical Methods in Quality Control*, Englewood Cliffs, NJ: Prentice-Hall, Inc. 1957.

Crosby, Phillip B. *Cutting the Cost of Quality*, Boston, Massachusetts: Industrial Education Institute, 1967.

Delbecq, A. L.; Van de Ven, A. H.; and Gustafson, D. H., *Group Techniques for Program Planning: A Guide to Nominal Group and Delphi Processes*, Glenview, IL: Scott, Foresman and Company, 1975.

Deming, W. Edwards. *Some Theory of Sampling*, New York: John Wiley & Sons, 1950.

Deming, W. Edwards, *Elementary Principles of The Statistical Control of Quality*, Second Printing June 1952. Tokyo: Nippon Kagaku Gijutsu Remmei, 1951.

Deming, W. Edwards, *Quality, Productivity and Competitive Position*, Cambridge, MA, Massachusetts Institute of Technology, Center for Advanced Engineering Study, 1982.

Dewar, Donald L., *The Quality Circle Guide to Participation Management*, Englewood Cliffs: Prentice-Hall, 1980.

Duncan, Acheson J. *Quality Control and Industrial Statistics*, 4th edition. Homewood, Illinois: Richard D. Irwin, Inc., 1965.

Dutschke, Wolfgang. *Qualitaetsregelung in der Fertigung*, Berlin, Germany: Springer Verlag, 1964.

Feigenbaum, Armand V., *Total Quality Control Engineering and Management*, revision originally published title: *Quality Control*, New York: McGraw-Hill, 1961.

Fetter, Robert B. *The Quality Control System*, Homewood, Illinois: Richard D. Irwin, Inc., 1967.

Grant, Eugene L. *Statistical Quality Control*, 3rd edition. New York: McGraw-Hill, 1964.

Groocock, J. M. *The Cost of Quality*, New York: Pitman Publishing, 1974.

Guerdan, Renee. *Byzantium, Its Triumph and Tragedy*, New York: George Allen, 1956.

Guide for Quality Control and Control Chart Method of Analyzing Data – ANSI Standard Z1.1-1969 and Z1.2-1969, New York: American National Standards Institute, Inc., 1969.

Hagan, John T. *A Management Role for Quality Control*, New York: American Management Association, Inc., 1968.

Hoffman, Thomas R. *Production Management and Manufacturing Systems*, Belmont, California: Wadsworth Publishing Company, 1967.

Juran, Joseph M. (Editor-in-Chief), *Quality Control Handbook*, 2nd edition. New York: McGraw-Hill Book Company, 1962.

Juran, Joseph M., *Managerial Breakthrough*, New York: McGraw-Hill, 1964.

Juran, J. M. and Gryna, F. M. Jr., *Quality Planning and Analysis*, Second Edition, New York: McGraw-Hill Book Company, 1980, p. 107.

Kirkpatrick, E. C. *Quality Control for Managers and Engineers*, New York: John Wiley & Sons, 1970.

Langevin, Roger C. *Quality Control in the Service Industry*, New York: AMACOM, 1977.

Latzko, William; Leszczynski, Walter; Morogiello, Anthony; O'Leary, Jack; and Sullivan, Kenneth. *MICR Quality Control Handbook*, Washington, D.C.: American Bankers Association, 1982.

Military Standard 105D, "Sampling Procedure and Tables for Inspection by Attributes," Washington, D.C.: U.S. G.P.O., 1963.

Morris, W. T. and The Ohio State University Productivity Research Group, *Measuring and Improving the Productivity of Administrative Computing and Information Services: A Manual of Structured, Participative Methods*, NSF Grant APR 75-20561.

Naus, Joseph I. *Data Quality Control and Editing*, New York: Marcel Dekker, Inc., 1975.

Nixon, Frank. *Managing to Achieve Quality and Reliability*, New York: McGraw-Hill Book Company, 1975.

Ott, Ellis R. *Process Quality Control*, New York: McGraw-Hill Book Company, 1975.

Peat, Marwick, Mitchell & Co. *Quality Controls Accounting and Auditing*, New York, 1976.

Price, Waterhouse & Co. *Quality Control in a Large Professional Practice*, 2nd edition. New York, 1975.

Print Specification for Magnetic Ink Character Recognition, X3.2-1970 and *Specification for Placement and Location of MICR Printing*, X9.13-1983, Amer. Nat. Stds. Inst., New York. Specification X3.2 is in process of revision by Comm. X9.

Roberts, Norman H. *Mathematical Methods in Reliability Engineering*, New York: McGraw-Hill Book Company, 1964.

Samsom, Charles; Hart, Philip; and Rubin, Charles. *Fundamentals of Statistical Quality Control*, Reading, Massachusetts: Addison-Wesley, 1970.

Sakardi, K. and Vincze, I. *Mathematical Methods of Statistical Quality Control*, Budapest, Hungary: New York/London Academic Press, 1974.

Schmidt, J. W. and Taylor, R.E. *Simulation and Analysis of Industrial Systems*, Homewood, Illinois: Richard D. Irwin, 1970.

Shewhart, Walter A. *Economic Control of Quality of Manufactured Product*, Princeton, New Jersey: D. Van Nostrand Company, 1931.

Shewhart, W. A. and Deming, E. W. *Statistical Method from the*

Viewpoint of Quality Control, Washington, D.C.: Department of Agriculture, 1939.

Simmons, David A. *Practical Quality Control*, Reading Massachusetts: Addison-Wesley, 1970.

Stafeil, Walter W. *1974 Survey of the Check Collection System*, Park Ridge, Illinois: Bank Administration Institute, 1975.

Supplement to the Common Machine Language for Mechanized Check Handling, (147R3), American Bankers Association, Washington, D.C., 1971.

The Common Machine Language for Mechanized Check Handling, Publication 147R3, American Bankers Association, New York, 1967.

Articles from Books

Bass, Leon. "Short-run Quality Control," in *Quality Control in Action*, Report Number 9. New York: American Management Association, 1958, p. 82.

Berger, Roger W. "Developing Quality Information Systems," in the *Administrative Applications Division of the American Society for Quality Control 1976 Yearbook*, Milwaukee, Wisconsin, 1975. p. 58.

Brown, A.W. "Professionalism—Let's Give It a New Dimension," in the *Administrative Applications Division of the American Society for Quality Control 1976 Yearbook*, Milwaukee, Wisconsin, 1975, p. 20.

Carter, C. L., Jr. "Results and How to Get Them," in the *Administrative Applications Division of the American Society for Quality Control 1976 Yearbook*, Hot Springs, Arkansas: S. G. Johnson, editor, 1976, pp. 49-55.

Day, Carl A. "What Management Can Expect from Quality Control," in *Quality Control in Action*, Report Number 9. American Management Association, 1958, p. 17.

Dertinger, E. F. "Quality Assurance: A New Organizational Concept," in the *New Concepts in Manufacturing Management*, New York: American Management Association, Inc., 1961, pp. 50-55.

Evans, Gordon H. "Manufacturing Staff Service," in *Managerial Job Descriptions in Manufacturing*, New York: American Management Association, Inc. 1964, pp. 249-80.

Feigenbaum, Armand V. "Total Quality Control," in *Quality Control in Action*, Report Number 9. American Management Association, 1958, p. 35.

Gadzinski, Chester. "Control of Quality Costs in Manufacturing," in Finley, R.E. and Ziobro, H. R. (Editors). *The Manufacturing Man and His Job*, New York: American Management Association, 1966, pp. 189-202.

Georgis, George S. "The Application of Quality Control in Purchasing-Vendor Relationships," in *Quality Control in Action*, Report Number 9. American Management Association, 1958, p. 90.

Goetz, Albrecht (Translator). "The Laws of Eshnunna," in Pritchard, James B. (Editor), *The Ancient Near East*, Vol I., 6th edition, 1973. Princeton, New Jersey: Princeton University Press, 1958.

Hallowell, H. Thomas, Jr. "Today's Management and Tomorrow's Products: The Need for Quality Control," in *Quality Control in Action*, Report Number 9. American Management Association, 1958, p. 7.

Holliday, W. A. "Management and the Use of Sampling Plans," in the *Administrative Applications Division of the American Society for Quality Control 1975 Yearbook*, Milwaukee, Wisconsin, 1975. p. 11.

Juran, Joseph M. "Identifying and Solving the Company's Major Quality Problems," in *Quality Control in Action*, Report Number 9. American Management Association, 1958, p. 27.

Juran, Joseph M., "Section 18 Motivation," in J. M. Juran, F. M. Gryna, Jr., and R. S. Bingham, Jr. (Editors), *Quality Control Handbook*, Third Edition, New York: McGraw-Hill Book Company, 1979.

Kirby, E. "Quality Control in Banking," in the *Administrative Applications Division of the American Society for Quality Control 1976 Yearbook*, Hot Springs, Arkansas: S. G. Johnson, editor, 1976, pp. 62-72.

Kruger, Henry. "The Visual Presentation of Data to Management," in *Men, Machines and Methods in the Modern Office, Report Number 6, New York: American Management Association, Inc.* 1958, pp. 56-90.

Latzko, William J. "Reduction of mistakes in a bank," in Deming, W. Edwards. *Quality, Productivity, and Competitive Position*, Boston, Massachusetts: Massachusetts Institute of Technology, Center for Advanced Engineering Study, 1982.

Lefevre, H. L. "Budgeting Time," in the *Administrative Applications Division of the American Society for Quality Control 1975 Yearbook*, Milwaukee, Wisconsin, 1975. p. 6.

McBride, Vernon. "Controlling the Cost of Quality," in the *Administrative Applications Division of the American Society for Quality Control 1976 Yearbook*, Hot Springs, Arkansas: S. G. Johnson, editor, 1976, pp. 20-26.

MacCrehan, William A., Jr. "Cost Considerations in Planning a Quality Control Program," in *Quality Control in Action*, Report number 9. New York: American Management Association, Inc., p. 48.

Marti, Carolyn M. "Mitigating Murphy's Law," in the *Administrative Applications Division of the American Society for Quality Control 1976 Yearbook*, Hot Springs, Arkansas: S. G. Johnson, editor, 1976, pp. 16-19.

Meck, F. "What Makes Incompetence in Management," in the *Administrative Applications Division of the American Society for Quality Control 1975 Yearbook*, Milwaukee, Wisconsin, 1975. p. 30.

Meek, Theophile J. (Translator). "The Code of Hammurabi," in Pritchard, James B. (Editor). *The Ancient Near East*, Vol. I., 6th edition 1973. Princeton, New Jersey: Princeton University Press, 1958, p.138ff.

Murdock, Bennett B. "Quality Control in Clerical Operations," in *Leadership in the Office*, New York: American Management Association, Inc. 1963, pp. 243-247.

Nadler, Paul S. "A Look at the Future in American Banking," in Prochnow, Herbert V. and Prochnow, Herbert V., Jr. (Editors). *The Changing World of Banking*, New York: Harper & Row, 1974, pp. 385-386.

Nick, Sam S. "Quality Control Problems — Case History Number 22," in the *Administrative Applications Division of the American Society for Quality Control 1976 Yearbook*, Hot Springs, Arkansas: S. G. Johnson, editor, 1976, pp. 62-72.

Rosander, A.C. "A Case Study in Controlling Quality & Products of a Large Nationwide Sample of Railroad Freight Traffic," in the *Administrative Applications Division of the American Society for Quality Control 1975 Yearbook*, Milwaukee, Wisconsin, 1975, p. 42.

Shaffir, Walter B. "Developing a Management Control 'Instrument Panel': A Practical Approach," in *Men, Machines and Methods in the Modern Office, Report Number 6*, New York: American Management Association, Inc. 1958, p. 61.

Sills, Budd. "Making Quality Control Work in the Office," in *Men, Machines and Methods in the Modern Office, Report Number 6*, New York: American Management Association, Inc. 1958, p. 72.

Sonderup, R. D. "Quality Control & Product Reliability," in Finley, R. H. and Ziobro, H. R. (Editors). *The Manufacturing Man and His Job*, New York: American Management Association, Inc. 1966, pp. 175-188.

Verigan, J. Frederick. "Motivating Operators and Supervisors to Better Quality," in *Quality Control in Action*, Report number 9. New York: American Management Association, Inc., 1958, p. 53.

Periodicals

Adam, Charles F. "Quality Control and the Challenge of Change," *Advanced Management Journal*, October 1966, pp. 7-12.

Akresh, Abraham. "Statistical Sampling: A Look at the Latest Developments," *Price, Waterhouse & Co, Staff News*, Vol. 16, No. 7 (September 1976), p. 5.

"American Society for Quality Control Standard A1 — Definitions, Symbols, Formulas and Tables for Control Charts," *Quality Progress*, October 1969, pp. 21-25.

"American Society for Quality Control Standard A2 — Definitions and

Symbols for Acceptance Sampling by Attributes," *Quality Progress*, September 1970, pp. 36-38.

Anderson, V. N. "Five Steps to Quality Control of Clerical Operations," *Systems and Procedures Journal*, November and December 1964, pp. 8-12.

Armstrong, G.R. "Producer's Quality Control Report Aids, Inspection of Incoming Parts," *Machinery*, April 1948, p. 1-4.

Bayer, Harmon S. and McElrath, Gayle W. "What Quality Control Needs From You," *Supervisory Management*, Vol 8, No. 11 (November 1963), pp. 10-13.

Beach, N. F. "Management and Quality Control," *Industrial Quality Control*, Vol. 22, No. 10 (August 1966), pp. 503-505.

Benz, William M. "Quality Control in the Office," *Industrial Quality Control*, Vol. 23, No. 11 (May 1967), pp. 531-534.

Bicking, Charles A. "Quality Control as a System," *Industrial Quality Control*, Vol. 23, No. 11 (May 1967), pp. 538-543.

Bicking, Charles A. "Cost and Value Aspects of Quality Control," *Industrial Quality Control*, Vol. 24, No. 6 (December 1967), pp. 306-308.

Buck, Vernon E. "Quality Control: Too Much Control — Too Little Quality?" *Management Review*, February 1966, pp. 30-34.

Budgell, Allston T. Jr., "The Manager's Stake in Quality Control," *Management Review*, November 1967, pp. 4-8.

Case, K. E.; Schmidt, J. W. and Bennett, G. K. "Cost Based Acceptance Sampling," *Industrial Engineering*, Vol. 4 (November 1972), p. 26.

Connell, F. M. Jr. "Statistical Quality Control of Clerical Operations," *Industrial Quality Control*, Vol. 24, No. 3 (September 1967), p. 154.

Cooke, Blaine. "The Pursuit of Perfection," *Industrial Quality Control*, Vol. 21, No. 1 (July 1964), pp. 16-20.

Cooke, Blaine. "Product Quality: The Pursuit of Perfection," *Management Review*, September 1964, pp. 21-24.

Deming, W. Edwards. "My View of Quality Control in Japan," *Report of Statistical Application Research*, Union of Japanese Scientists and Engineers, Vol. 22, No. 2 (June 1975) pp. 7-8.

Deming, W. Edwards. "On Some Statistical Aids Toward Economic Production," *Interfaces*, Vol. 5, Revised December 7, 1973, (September 4, 1973).

Deming, W. Edwards. "Report to Management," *Industrial Quality Control*, Vol. VI, 1972.

Deming, W. Edwards and Hansen, Morris H. "Some Theory on the Influence of the Inspector and Environmental Conditions, With an Example," *Statisticia Neerlandica*, Vol. 26, No. 3, 1972, pp. 101-112.

Eldridge, Lawrence A.; and Aubrey, Charles A. II, "Stressing Quality — The Path to Productivity," *The Magazine of Bank Administration*, Volume 57, Number 6, June 1983, p. 22.

Enrick, Norbert Lloyd. "Control Charts for Inventory," *Quality Assurance*, December 1967, pp. 24-27.

Exton, William, Jr. "How to Improve Clerical Accuracy," *Supervisory Management*, April 1971, p. 30 (Condensed from *Personnel Journal, Inc.*, Vol. 49, No. 8, 1970).

Exton, William, Jr. "How error-prone is your bank?" *Banking*, May, 1977.

Fairman, J. R. "Bonus Plan Rewards Work of Mail and Copying Teams," *Administrative Management Association*, Inc., February 1967, p. 57.

Fiddler, T. R. "Quality is Essential for Successful Merchandizing," *Stores*, December 1966, pp. 32-33.

Field, David L. "Flinching — A Factor in Estimating Success Probabilities," *Industrial Quality Control*, Vol. 21, No. 8, February 1965, pp. 406-408.

Field, David L. "Thoughts on the Economics of Quality," *Industrial Quality Control*, Vol. 23, No. 4, October 1966, pp. 178-184.

Fiorenza, Frank A. "Accounting for Unusual Quality Control Costs," *Management Accounting*, February 1969, pp. 53-55.

Flannery, Patrick M. "Keep Raising Your Employee's Goals," *Administrative Management Society*, February 1967, pp. 57-58.

General Professional Council of the Administrative Applications Division of the American Society for Quality Control. "The Basic Work Elements of Quality Control Engineering," *Industrial Quality Control*, May 1961, pp. 8-10.

Golomski, William A. "Are You Selling Quality Short?" *Nation's Business*, December 1967, pp. 72-74.

Golomski, William A. "Quality Control — History in the Making," *Quality Progress*, Vol. 9, No. 7, July 1976, pp. 16-18.

Goodfellow, Matthew, "Quality Control Circle Programs — What

Works and What Doesn't," *Quality Progress*, Volume 19, Number 8 (August 1981), p. 30.

Gryna, Frank, Jr. "User Quality Costs," *Quality Progress*, Vol. V, No. II, November 1972, p. 18.

Halpern, J. "Good Enough Is Not Enough," *Personnel Journal*, Vol. 45, No. 621, November 1966, p. 2.

Harding, Harry C. "Best Cost Sampling System for Purchased Components," *Journal of Purchasing*, February 1968, p. 29-42.

Heil, Joseph B., Jr. "Using 'Sampling Methods' for Quality Control," *Administrative Management Society*, February 1967, pp. 55-56.

Heim, Arthur I., Jr. "Cut Costs with Quality Control," *Supervisory Management*, Vol. 4, No. 1, January 1959, pp. 55-56.

Huff, Darrell. "*Parlez-Vouz* Statistics?" *Chemical Division News*, (Reprinted from *Think*, magazine, 1963), June 1965, pp. 3-6.

Jacobson, Henry J. "Management Methods of Inspection Control," *Industrial Quality Control*, Vol. 21, No. 1 , July 1964, pp. 24-28.

Juran, J. M. "Activities and Labels; Functions and Names," *Industrial Quality Control*, Vol. 24, No. 5, November 1967, pp. 248-250.

Juran, J. M. "10 Basic Tools to Control Your Vendor's Quality," *Factory Management & Maintenance*, Vol. 113, No. 3, March 1955, pp. 126-129.

Juran, J. M. "Can Your Processes Hold the Tolerances You've Set?" *Factory Management*, Vol. 110, No. 6, June 1952, pp. 118-120.

Juran, J. M. "Insure Success for your Quality Control Program," *Factory Management & Maintenance*, Vol. 108, No. 10, October 1950, pp. 106-109.

Juran, J. M. "Is Your Product Too Busy?" *Factory Management & Maintenance*, Vol, 110, No. 8, August 1952, pp. 125-128.

Juran, J. M. "Nine Steps to Better Quality," *Factory Management & Maintenance*, Vol 110, No. 11, November 1952, pp. 106-108.

Juran, J. M. "The Quality Control Circle Phenomenon," *Industrial Quality Control*, Vol. 23, No. 7, January 1967, pp. 329-336.

Juran, J. M. "Quality Control of Service — The 1974 Japanese Symposium," *Quality Progress*, Vol. 8, No. 4, April 1975, pp. 10-13.

Juran, J. M. "Quality Problems, Remedies and Nostrums," *Industrial Quality Control*, Vol. 22, No. 12, June 1966.

Juran, J. M. "Six Steps to Inspection Management," *Factory Management & Maintenance*, Vol. 110, No. 1, November 1952, pp. 114-117.

Juran, J. M. "The Two Worlds of Quality Control," *Industrial Quality Control*, Vol. 2, No. 5, November 1964, pp. 238-244.

Latzko, William J. "A Quality Control System For Banks," *The Magazine of Bank Administration*, Vol. XLVII, No. 11 November 1972, pp. 17-23.

Latzko, William J. "Quality Control For Banks," *The Bankers Magazine*, Vol. 160, No. 4, Autumn 1977, pp. 64-68.

Latzko, William J. "Why Banks Need Bullseye Accuracy and Consistency," *Printing Impressions*, Vol. 20, No. 7, December 1977, pp. 10J-10O.

Latzko, William J., "Quality will determine winners in competitive banking market," *Bank System & Equipment*, Vol 21, No. 5, May 1984 p. 204.

Latzko, William J., "Quality Circles Won't Work Without Quality Control," *The Magazine of Bank Administration*, Volume 55, Number 12, December 1981, p. 23.

Latzko, William J. "The Paperwork Factory," *Quality*, Vol. 23, No. 3, March 1984, pp. 31-33.

Latzko, William J. "Process Capability in Administrative Applications," *Quality Progress*, Vol. 18, No. 6, June 1985, pp. 70-73.

Lieberman, William L. "Organization and Administration of a Quality Control Program," *Industrial Quality Control*, January 1962, pp. 27-30.

Main, Jeremy, "The curmudgeon who talks tough on quality," *Fortune*, June 25, 1984, p. 119.

Marash, Stanley A. "Performing Quality Audits," *Industrial Quality Control*, January 1966, pp. 342-347.

McCreary, Robert M. "Whence Cometh Quality Control?" *Quality Progress*, Vol 9, No. 7 (July 1976), p. 15.

Medlin, John (Assoc. Editor). "New Controls Help Insure Quality," *Administrative Management*, October 1967, pp. 20-26.

Melan, E. H., "Process Management in Service and Administrative Operations," *Quality Progress*, June 1985.

Midlam, K. D. II, "From the Scrap Box," *Industrial Quality Control*, January 1962, p. 30.

Mosteller, Frederick and Tukey, John W. "The Uses and Usefulness of Binomial Probability Paper," *American Statistical Association Journal*, June 1949, pp. 172-212.

Namias, Jean. " A Rapid Method to Detect Differences in Interviewer Performance," *Journal of Marketing*, Vol 26, No. 2 (April 1962), pp. 68-72.

Nelson, Ronald. "All-Out Quality Control Slashes Customer Rejections," *Steel*, July 1968, pp. 49-50.

Newman, Benjamin. "Management Controls," Peat, Marwick, Mitchell & Co. Bulletin, Vol 13, No 5 (May 1966).

Pabst, William R., Jr. "Motivating People in Japan," *Quality Progress*, Vol. 10 (October 1972), pp. 14-18.

Palmer, Barry A. "Quality Control Engineering," *Industrial Quality Control*, May 1964, pp. 17-18.

Prabhu, P. R. "Consumers, Value & Management," *Quality Progress*, Vol 8, No 4 (April 1975), pp. 8-9.

Prevete, Joan. "Citybank's Trace System Helps Cut Reject Volume in Half," *Bank Systems and Equipment News*, May 1975, pp. 38-39.

Richman, Alan. "Bankers Trust Combats Rejects With Strong Prevention Program," *Bank Systems and Equipment News*, September 1975, pp. 54-57.

Richman, Alan. "First Pennsylvania Bank Marries On-Line Reconcilement to Physical Handling of MICR Rejects," *Bank Systems and Equipment News*, June 1975, pp. 54-57.

Richman, Alan. "PNB Reject System Eases Transit Bottleneck, Speeds Funds," *Bank Systems and Equipment News*, August 1975, pp. 51-53.

Rosander, A. C., Guterman, H. E., and McKeon, A. J. "The Use of Random Work Sampling for Cost Analysis & Control," *American Statistical Association Journal*, 1959, pp. 382-397.

Sandholm, Lennart, "Japanese Quality Circles—A Remedy for the West's Quality Problems?" *Quality Progress*, Volume XVI (February 1983), p. 21

Sheldon, George W. and Finch, Frederick E. "Bank Queues: A Comparative Analysis of Waiting Lines," *The Magazine of Bank Administration*, July 1976, pp. 31-35.

Smith, Martin. "Quality Control in The Office," *Journal of Systems Management*, April 1974, pp. 34-36.

Squires, Frank H. "Product Quality: Pretty Good Isn't Good Enough," *The Management Review*, Vol 48, No 10 (October 1959), pp. 18-23.

Surles, Lynn and Stanbury, W. A., Jr. "Building Quality Into Employee Job Habits," *Supervisory Management*, April 1965, pp. 14-20.

Taylor, J. E. "Statistical Quality Control," *Administrative Management Society*, April 1965, pp. 14-20.

Vance, L. L. "Glossary of Statistical Terms for Accountants," *American Institute of Certified Public Accountants*, September 8, pp. 1-30.

Newspapers

"Advertisement Supplement," *American Banker*, November 15, 1976.

"BAI Exception Item Handling Conference Invites Industry-wide Effort for Control," *American Banker*, March 17, 1976, p. 6.

"Careless Printing Costly to Banks," *Washington Letter*, March 8, 1976.

DeMott, J. S., "Stop Sign," *Time*, (New York) March 11, 1985 p. 48

"Dubuque Bank Gets a Little Too Friendly," New York *Times*, November 14, 1976, p. 14.

"Poll Finds Consumers Unhappy on Quality," *American Banker*, September 22, 1976, p. 7.

Stafeil, Walter, W. "Exception Item 'Horror Story' May Yet Have a Happy Ending," *American Banker*, May 16, 1977.

"Trends," *The Reporter Dispatch*, (White Plains), May 16, 1976.

Pamphlets and Published Reports

Administrative Applications Division of the American Society for Quality Control. "General Requirements for a Quality Control

Program," *Specification of General Requirements for a Quality Program*, Milwaukee, Wisconsin, 1968.

Bank Check Specification for MICR, X3.3-1970, American Standards Institute, New York, New York, 1970.

Deming, W. Edwards. "A Career in Statistical Administration," *New York University*, August 1953.

Deming, W. Edwards. "Probability as a Basis for Action," Number 145, a private publication, November 12, 1973.

ITT Guidebook for Their Management. "Quality Improvement Through Defect Prevention," 1967.

National Industrial Conference Board, Inc. "Quality Control Methods and Company Plans (Report)," *Conference Board Report*, Studies in Business Policy, Vol 36, May 1949.

Print Specification for Magnetic Ink Character Recognition, X3.2-1970, American National Standards Institute, New York, 1970.

Stafeil, Walter. "Exception Item Handling project," *Bank Administration Institute*, March 27, 1973.

Supplement to the Common Machine Language for Mechanized Check Handling, (147R3), American Bankers Association, Washington, D.C., 1971.

The Common Machine Language for Mechanized Check Handling, Publication 147R3, American Bankers Association, Washington, D.C., 1956.

U. S. Department of Agriculture Consumer and Marketing Services Statistical Staff. "Sampling Facts," January 1970.

Transactions

Abrahams, Ralph G. "Economic Aspects of Quality Control," in *Twenty-Fourth Annual Conference Proceedings*, (New Brunswick: Metropolitan Section, American Society for Quality Control, 1972), pp. 1 ff.

Agnone, Anthony M., Brewer, Clayton C., & Caine, V. "Quality Cost Measurement and Control," in *Twenty-Seventh Annual Conference Proceedings*, (Milwaukee: American Society for Quality Control, 1973), pp. 300-306.

Barone, Robert E. and Harkness, Donald R. "Innovations in Software Quality Management," in *Twenty-Seventh Annual Conference Proceedings*, (Milwaukee: American Society for Quality Control, 1973), pp. 152-156.

Churchill, G. W. "Minimizing the Cost of Lot Sampling Through Solution of Cost-Probability Equations," in *Twenty-Second Annual Conference Proceedings*, (Philadelphia: American Society for Quality Control, 1968), pp. 643 ff.

Cole, Robert E., "Common Misconceptions of Japanese QC Circles," in *Thirty-Fifth Annual Quality Congress Transactions*, (San Francisco: American Society for Quality Control, 1981), p. 188-189.

Combes, James; Klaus, George and Roberta, Frank. "Management View of Quality Cost," in *Twenty-Seventh Annual Conference Proceedings*, (Milwaukee: American Society for Quality Control, 1973).

Dawes, Edgar W. "Optimizing Attribute Sampling Cost," in *Twenty-Seventh Annual Conference Proceedings*, (Milwaukee: American Society for Quality Control, 1973), pp. 181-187.

Deming, W. Edwards. "Some Statistical Logic in the Management of Quality," in *All India Conference on Quality Control Proceedings*, (New Delhi, May 1971), pp. 98 ff.

Ekvall, D. N. "Measuring the Profitability of Quality Control Effectiveness," in *Twenty-Sixth Annual Conference Proceedings*, (Milwaukee: American Society for Quality Control, 1972), pp. 219-222.

Fasteau, Herman H., Ingram, Jack J., and Minton, George. "Control of Quality of Coding in the 1960 Censuses," in *Mid-Atlantic Conference Proceedings*, (Washington, 1962), pp. 1-24.

Fasteau, Herman H., Ingram, Jack J., and Mills, Ruth H. " Study of the Reliability of Coding of Census Returns," in *Sixteenth Annual Conference Proceedings*, (Minneapolis: American Society for Quality Control, 1967).

Fitzgerald, J. P. "Managing the Total Quality System," in *Twenty-Eighth Annual Conference Proceedings*, (Middlesex, NJ: Metropolitan Section, American Society for Quality Control, 1976).

Gaugler, David L. "Controlling Quality at an Internal Revenue Center," in *Twenty-Ninth Annual Conference Proceedings*,

(San Diego: American Society for Quality Control, 1975), pp. 228 ff.

Gonet, John J. "Improving the Management of Quality Cost," in *Twenty-Second Annual Conference Proceedings*, (Philadelphia: American Society for Quality Control, 1968), pp. 261 ff.

Gookins, Bud. "Organizing for Quality," in *Twenty-Second Annual Conference Proceedings*, (Philadelphia: American Society for Quality Control, 1968), pp. 709 ff.

Hansen, M. H.; Fasteau, H. H.; Ingram, J. J. and Minton, G. "Quality Control in the 1960 Censuses," in *Proceedings of 1962 Middle Atlantic Conference*, (Washington: American Society for Quality Control, 1973), pp. 311 ff.

Hagan, John T. "Quality Costs at Work," in *Twenty-Seventh Annual Conference Proceedings*, (Milwaukee: American Society for Quality Control, 1973), pp. 37 ff.

Hill, Hubert M. "A Systems Definition of Total Quality Control," in *Twenty-Seventh Annual Conference Proceedings*, (Milwaukee: American Society for Quality Control, 1973), pp. 35-40.

Hunter, Thomas M. "A Computer Evaluation of Toll Collectors," in *Eighteenth Annual Conference Proceedings*, (Buffalo: American Society for Quality Control, 1964), pp. 288 ff.

Johnson, Lynwood A. "Quality and Productivity Control of Mail Preparation Operations," in *Twenty-Ninth Annual Conference Proceedings*, (San Diego: American Society for Quality Control, 1975), pp. 282 ff.

Kirby, Eugene. "Quality Control in Banking," in *Twenty-Ninth Annual Conference Proceedings*, (San Diego: American Society for Quality Control, 1975), pp. 258 ff.

Krehbeiel, W. R. "Customer Oriented Quality Definitions," in *Eighteenth Annual Conference Proceedings*, (Buffalo: American Society for Quality Control, 1964), pp. 299 ff.

Langevin, Roger C. "General Quality Control Model for Bank Operations," in *Twenty-Fifth Annual Conference Proceedings*, (New Brunswick: Metropolitan Section, American Society for Quality Control, 1973).

Latzko, William J. "Stabilized t-Charts: Theory and Practice," in *Twenty-Third Annual Conference Proceedings*, (Los Angeles: American Society for Quality Control, 1969), pp. 657 ff.

Latzko, William J. "Quality Control in Banking," in *Twenty-Fourth Annual Conference Proceedings*, (New Brunswick: Metropolitan Section, American Society for Quality Control, 1972), pp. 61 ff.

Latzko, William J. "Clerical Process Capability," in *Twenty-Fifth Annual Conference Proceedings*, (New Brunswick: Metropolitan Section, American Society for Quality Control, 1973), pp. 131 ff.

Latzko, William J., "Quality Control of MICR Input," in *Twelfth Annual Conference Proceedings*, (Hempstead, NY: Long Island Section, American Society for Quality Control, 1974) pp. 38 ff. This paper details the specifications.

Latzko, William J. "Quality Control in Banking," *National Operations and Automation Conference Proceedings*, (New York: American Bankers Association, 1974), pp. 36-49.

Latzko, William J. "Reducing Clerical Quality Costs," in *Twenty-Eighth Annual Conference Proceedings*, (Boston: American Society for Quality Control, 1974), pp. 185 ff.

Latzko, William J. "QUIP — The Quality Improvement Program," in *Twenty-Ninth Annual Conference Proceedings*, (San Diego: American Society for Quality Control, 1975), pp. 246 ff.

Latzko, William J. "Basic Tools for Clerical Quality Control," in *Twenty-Seventh Annual Conference Proceedings*, (New Brunswick: Metropolitan Section, American Society for Quality Control, 1975), pp. 27 ff.

Latzko, William J., "Statistical Quality Control of MICR Documents", in *Thirty-First Annual Technical Conference Transactions*, (Philadelphia, PA: American Society for Quality Control, 1977), p. 118.

Latzko, William J., "Quality Productivity Measures—Participative Management," *Thirty-Fifth Annual Quality Congress Transactions*, (San Francisco: American Society for Quality Control, 1981), p. 389-395.

Latzko, William J.,"Minimizing the Cost of Inspection," in *Thirty-Sixth Annual Quality Congress Proceedings*, (Detroit: ASQC 1982).

Lieberman, William L. "Analysis of Two Modes of Quality Program Failures," in *Twenty-Second Annual Conference Proceedings*,

(Philadelphia: American Society for Quality Control, 1968), pp. 245 ff.

Mundel, August B. "Organizing to Achieve Quality," in *Twenty-Fifth Annual Conference Proceedings*, (New Brunswick: Metropolitan Section, American Society for Quality Control, 1973), pp. 80 ff.

Mundel, August B. "Quality Control in Non-Hardware Applications," in *Twenty-Fifth Annual Conference Proceedings*, (Milwaukee: American Society for Quality Control, 1971), pp. 29-36.

O'Brien, James L. "Some Promising Approaches to Computerizing Administrative Operations," in *Eighteenth Annual Conference Proceedings*, (Buffalo: American Society for Quality Control, 1964), pp. 1-22.

Olmstead, Blair E. "Quality Control Applied to Clerical Operations," in *Twenty-Second Annual Conference Proceedings*, (New Brunswick: Metropolitan Section, American Society for Quality Control, 1970).

Ott, Ellis R.; Frey, William C.; and Schecter, Edwin S. "Statistical Quality Control: Some Basic Concepts," in *Twenty-Third Annual Conference Proceedings*, (New Brunswick: Metropolitan Sect., American Society for Quality Control, 1970), pp. 1.

Robert, Paul A. "Quality Control Management at the Corporate Level," in *Twenty-First Annual Conference Proceedings*, (Chicago: American Society for Quality Control, 1967), pp. 3 ff.

Saggese, Richard. "Quality Control, Its Day Has Come," in *Twelfth Annual Conference Proceedings*, (Hempstead, LI: Long Island Section, American Society for Quality Control, 1974), pp. 37 ff.

Schock, Harvey E., Jr. "What Management Expects from a Quality Assurance System," in *Twenty-Seventh Annual Conference Proceedings*, (New Brunswick: Metropolitan Section, American Society for Quality Control, 1975), pp. 61 ff.

Staab, Thomas C. "Quality Applicable to Paperwork? — Probably!" in *Twenty-Seventh Annual Conference Proceedings*, (Milwaukee: American Society for Quality Control, 1973), pp. 393-397.

Trap, Brian E. "The Building of an Effective MICR Quality Control

Program," in *Twenty-Ninth Annual Conference Proceedings*, (San Diego: American Society for Quality Control, 1975), pp. 255 ff.

Vincent, Robert W. "Quality Cost Administration," in *Twenty-Third Annual Conference Proceedings*, (New Brunswick: Metropolitan Section, American Society for Quality Control, 1971), pp. 1-5.

Theses

Alden, Robert E. *Clerical Work Measurement: A Management Tool*, New Brunswick, New Jersey: Stonier Graduate School of Banking, 1971.

Chappas, Charles H. *System Planning — The Key for Effective Bank Management*, New Brunswick, New Jersey: Stonier Graduate School of Banking, 1968.

Denny, James E. *Problem Solving in the Bank*, New Brunswick, New Jersey: Stonier Graduate School of Banking, 1974.

Hammer, Frederick S. *Management Science in Banking*, New Brunswick, New Jersey: Stonier Graduate School of Banking, 1966.

Kolberg, Harvey C. *The Application of Systems Principles and Technology to Bank Operations in Small and Medium Size Banks*, New Brunswick, New Jersey: Stonier Graduate School of Banking, 1966.

Murray, Malcolm T., Jr. *Loan Quality Control and Branch Banking*, New Brunswick, New Jersey: Stonier Graduate School of Banking, 1975.

Reed, Clarence R. *Evaluating the Performance of Commercial Loan Officers*, New Brunswick, New Jersey: Stonier Graduate School of Banking, 1969.

Government Publications

Bingham M. D. on *1963 Economic Censuses: Verification of Punching of Data Cards from 1963 Economic Censuses Reports*, (Washington: U. S. Bureau of Census, February 5, 1964).

Cook, William H. *Sample Verification Plan for Punching CATO*

Cards, (Washington: U. S. Bureau of Census, November 5, 1957), pp. 1-8.

Ingram, Jack J. *Quality Control of Clerical Operations*, (Washington: U. S. Bureau of Census, November 5, 1962), pp. 1-48.

United States Bureau of Census. *Memorandum on Sample Verification and Quality Control Methods in the 1950 Census*, (Washington, D.C.: 1963).

United States Department of Defense. *Military Standard 105D. Sampling Procedures and Tables for Inspection by Attributes*, (MIL-STD-105D). (Washington, D.C.: U.S. Government Printing Office, 1963).

United States Department of Defense. *Military Standard 414, Sampling Procedures and Tables for Inspection by Variables for Percent Defective*, (Washington, D.C.: U.S. Government Printing Office, 1957).

United States Department of Defense. *Military Standard 9858A Quality Program Requirements (MIL-Q-9858A)*, (Washington, D.C.: U.S. Government Printing Office, 1963).

Speeches

Deming, W. Edwards. “On Some New Principles in Administration,” a speech presented at the All Day Joint Conference of the Metropolitan Sections of the American Society for Quality Control and the American Statistical Association, May 15, 1976.

Ishikawa, Kaoru, “Japanese Total Quality Control,” part of a panel discussion presented at the American Society for Quality Control’s 36th Annual Quality Congress, Detroit Michigan, May 4, 1982.

Latzko, William J. “The MICR Challenge for Bankers,” a speech presented at the Bank Administration Institute’s Exception Item Conference, Chicago March 1976.

Latzko, William J., “Dr. Deming’s 14 Points” a speech presented at a meeting of the Stamford, Ct. Section of the American Society for Quality Control, January 9, 1985.

Latzko, William J., “Deming’s 14 Points in the Service Industry,” a speech presented at the annual conference of the Administrative Applications Division of the American Society for Quality Control, Williamburg, Va., March 21, 1985.

Monks, Donald R. "MICR Standards" a speech presented at the Bank Administration Institute's "MICR Document Testing Workshop", Baltimore MD, December 13-15, 1984.

Quinn Michael P. A lecture on organizing a quality control department in a bank presented at the American Institute of Banking, New York Chapter on January 2, 1978.

Schweitzer, Frederick A. "Quality Control in Direct Mail Operations," a speech presented at the Second Annual Mini-Seminar of the Metropolitan Section of the American Society for Quality Control, 1974.

Tribus, Myron, "Reducing Deming's 14 Points to Practice," an unpublished report enclosed in a letter from Dr. Myron Tribus, Director CAES MIT, June 13, 1983, pp. 12-13

Unpublished reports

Kirby, Eugene. "Results of a Recent Survey of Banks," unpublished report to the Banking Subcommittee of the Administrative Applications Division of the American Society for Quality Control, 1975. A copy of the report is enclosed in this book.

Tribus, Myron, "Reducing Deming's 14 Points to Practice," an unpublished report enclosed in a letter from Dr. Myron Tribus, Director CAES MIT, June 13, 1983, pp. 12-13.

Index